Emotions and the Right Side of the Brain

Guido Gainotti

Emotions and the Right Side of the Brain

 Springer

Guido Gainotti
Department of Neurosciences
Catholic University of the Sacred Heart Rome
Rome
Italy

ISBN 978-3-030-34092-6   ISBN 978-3-030-34090-2   (eBook)
https://doi.org/10.1007/978-3-030-34090-2

This Springer imprint is published by the registered company Springer Nature Switzerland AG
The registered company address is: Gewerbestrasse 11, 6330 Cham, Switzerland

Contents

Introduction

Contents

Approximately 50 years have elapsed since Terzian (1964), Rossi and Rosadini (1967) and Gainotti (1969, 1972) published the first papers on laterality effects in the representation of emotions that had a clear impact on subsequent investigations of this subject. According to Google Scholar, the number of papers dealing with this topic has steadily increased and has doubled every 10 years. Indeed, the notion of right hemisphere dominance for emotions or, alternatively, of different lateralisation of positive and negative emotions has become very popular. Different models of emotional lateralisation have been advanced since the first clinical observations raised this question and the impact of these models has been related to the research paradigms that were prevalent in the corresponding periods of time. Thus, the hypothesis of general dominance of the right hemisphere for all kinds of emotions was very popular during the first period of study, which was characterised by the prevalence of clinical and poorly theoretically motivated experimental investigations. On the other hand, the model of a different emotional specialisation of the right and left hemisphere (usually known as the 'valence hypothesis') gained influence with the development of investigations that studied comprehension and expression of positive and negative emotions in normal subjects. This model became even more influential when it was reframed by Davidson (1983) in terms of 'approach vs withdrawal' motivational tendencies. In more recent years, very interesting results were obtained by anatomo-clinical and activation studies that investigated laterality effects within specific brain structures (such as the amygdala, the ventromedial prefrontal cortex and the anterior insula), known to play a critical role in different

© Springer Nature Switzerland AG 2020
G. Gainotti, *Emotions and the Right Side of the Brain*,
https://doi.org/10.1007/978-3-030-34090-2_1

components of emotions. Furthermore, interesting information about the role of the right hemisphere in human emotions were provided by the study of behavioural and emotional disorders of patients with asymmetrical forms of fronto-temporal degeneration. In the present volume, the presentation and discussion of data relevant to understanding the relationships between emotions and brain laterality will be preceded, in Chap. 2, by an attempt to define the meaning of the term 'emotions' and to analyse similarities and differences between the two systems provided with adaptive value (i.e. the emotional and the cognitive system) which, in different situations, allow us to face a rapidly changing milieu.

In Chap. 3, attention will be shifted from the analysis of emotions to that of the brain structures that have a critical role in different components and hierarchical levels of emotions. Following these introductory sections, in Chap. 4, the history of research on emotional laterality, will be taken into account. This will be accomplished by considering separately: (a) the first descriptions of the different emotional behaviour shown by right and left brain-damaged patients; (b) the experimental and clinical investigations that have studied the non-verbal communicative aspects of emotions and (c) the experimental and clinical investigations that have studied the autonomic components of emotions. This historical part will be followed in Chap. 5 by a detailed analysis of recent trends in the study of the links between emotions and brain laterality. This will be carried out by considering separately: (I) the study of laterality effects in brain structures that have a critical role in different components of emotions and (II) the emotional and behavioural disorders of patients with asymmetrical forms of fronto-temporal degeneration. In Chap. 6, two general neurobiological models, which were advanced to explain results obtained following these recent lines of research, will be discussed. In Chap. 7, some psychopathological implications of the hemispheric asymmetries for emotions will be taken into account. The conclusion will be that no real conflict exists between models that surmise a right hemisphere dominance for emotions and those that hypothesise a different hemispheric representation of sympathetic and parasympathetic sections of the autonomic nervous system, because special relations exist between emotional system and sympathetic activities.

References

Davidson RJ. Hemispheric specialization for cognition and affect. In: Gale A, Edwards J, editors. Physiological correlates of human behavior. London: Academic Press; 1983. p. 203–26.

Gainotti G. Réaction catastrophiques et manifestations d'indifférence au cours des atteintes cérébrales. Neuropsychologia. 1969;7:195–204.

Gainotti G. Emotional behavior and hemispheric side of the lesion. Cortex. 1972;8:41–55.

Rossi GF, Rosadini G. Experimental analysis of cerebral dominance in man. In: Millikan CJ, Darly FL, editors. Brain mechanisms underlying speech and language. New York: Grune and Stratton; 1967.

Terzian H. Behavioural and EEG effects of intracarotid sodium amytal injection. Acta Neurochir. 1964;12:230–9.

What Are Emotions

2

Contents

2.1 Attempts to Define the Word 'Emotions' by Stressing Their Adaptive Value

In our daily life we are submerged by messages, coming from media and publicity that use the term 'emotions' in relation to different situations, such as cars, money, travels, exhibitions, football, clothing and chocolate, to promise us new and exciting experiences. This undifferentiated and misleading use of the word 'emotions' has forced researchers to give a more precise definition of this term, identifying the semantic traits that define and distinguish it from other similar words. There is not, however, a general consensus among scholars about the definition of this word and the boundaries of the concept of 'emotions' are rather vague and indistinct.

Most authors agree that emotions are rather complex and stereotyped behavioural schemata, characterised by particular types of subjective experience and by an increased activation of the vegetative system. However, there is much less agreement over which behavioural schemata to include in this area and how to demarcate emotions from other behavioural patterns belonging to contiguous but different areas. For instance, two terms such as 'anger' and 'love,' commonly considered as typical examples of emotions in the context of the current meaning of this word, have a different status for theorists of emotions, because anger is commonly considered one of the 'basic emotions,' whereas love is not included in this area. A rough but effective method of differentiating between these more or less similar

© Springer Nature Switzerland AG 2020
G. Gainotti, *Emotions and the Right Side of the Brain*,
https://doi.org/10.1007/978-3-030-34090-2_2

behavioural patterns could be to identify reference axes and in relation to them to distinguish the schemata that are part of the area of emotions from those that are not. A first reference axis could be the complexity and level of phylogenetic development of the behavioural schema in question. This axis could extend from the stage of simple reflexes to that of cognitive behaviour. On this axis, the emotions would be located at an intermediate level of complexity. On one side they would be differentiated by reflex behaviours (such as sneezing or flinching), which are very simple, primitive and innate behavioural schemata and, on the other side, by cognitive behaviours, with respect to which they are distinguished by lesser complexity due to their automatic (rather than intentional) nature and to the fact that they are anchored to relatively fixed response patterns.

A second reference axis might be based on the duration of the behavioural schema in question. This axis could range from very brief reactions, which are typically triggered by external events (as are the reflex responses we have just mentioned) to long-lasting behavioural schemata, which are stable over time and are not provoked by external stimuli. According to Ekman (1984), who attempted to analyse the problem with reference to this axis, emotions are reactions that last some seconds and are triggered by situational antecedents typical of each emotion (such as a threatening event for the emotion of fear or having received a prize for the emotion of happiness). They must, therefore, be distinguished by long-lasting affective schemata that are not usually related to external events, such as affect or personality traits. It must be acknowledged, however, that although the distinction between short-lasting emotional reactions and long-lasting affect and personality traits is very useful from a theoretical point of view, it should not be overemphasised. Several authors, such as Plutchik (1980), Ekman (1984) and Scherer (2000) acknowledged the close familiarity between emotions, affect and personality traits and agreed that affect is closely related to emotions, but refers to more enduring and less reactive states.

Another characteristic of emotions that most authors (e.g. Leventhal 1979, 1987; Scherer 1984; Oatley and Johnson-Laird 1987, 2014; Gainotti 1989, 2000; Montag and Panksepp 2017) consider typical of these behavioural schemata is their adaptive value. In fact, these authors consider emotions as phylogenetically advanced adaptive response patterns, which are based on the integrated activity of several components and have high survival value. Due to their complex structure, multicomponential nature and general adaptive functions, emotions form a truly adaptive system with functional architecture similar to that of the cognitive system, but with different goals.

2.2   Similarities and Differences Between the Emotional and the Cognitive System

Following this line of thought, Oatley and Johnson-Laird (1987, 2014) proposed that to face a partially unpredictable environment and to select the most appropriate plan of action from among those available, the organism has two operative sets of mechanisms that work together, i.e. the emotional and the cognitive system. The emotional system is considered an emergency system that is able to interrupt an ongoing action with an urgency procedure and to rapidly select a new operative schema. On the other hand, the cognitive system is considered a more evolved and advanced adaptive system. This system is capable of exhaustively analysing complex situations and working out new plastic plans that take into account both the external situation and the outcome of previous similar conditions but requires much more time to be put into action. Important similarities and differences exist between these two adaptive systems: the former are in the foreground from a structural point of view, and the latter emerge when we consider the goals of these two systems.

The structural similarity between emotional and cognitive system stems from the fact that the main components of these systems have the common functions of: (a) scanning the external milieu, focusing attention on the most relevant stimuli; (b) analysing these information to compute their meaning; (c) providing an appropriate, adaptive response; (d) memorizing the most relevant data (stimulus characteristics, subject's response and outcome of this response) inserting them into appropriate learning systems.

On the other hand, the differences between the two systems derive from the fact that the logic of an emergency network is different from that of a controlled modus operandi and that their components must, therefore, have partly different characteristics. The qualitative features that, according to most authors, characterise the main components of the emotional system are summarised in Table 2.1, where they are contrasted with the characteristics of the homologous components of the cognitive system.

Table 2.1 Main characteristics of the various components of the emotional and the cognitive system

Common function	Manner in which this function is accomplished by the emotional system	Manner in which this function is accomplished by the cognitive system
1. Orienting of attention	Automatic orientation elicited by external stimuli	Intentional, controlled orientation partly determined by internal representations
2. Analysis of sensory data	Quick computation of poorly processed sensory data, that leads to a direct motor response	Exhaustive but slower computation of highly processed information, that can lead to gather further information about the stimulus
3. Computation of stimulus significance	Concerns subjective, self-referential meaning	Concerns objective, social meaning
4. Response activated by the	Is immediately selected from a small number of innate operative patterns, which are endowed with high priority, include expressive components and concern both the motor and the vegetative system	Consists of a very large number of controlled and plastic strategic plans, which do not require a concomitant activation of the vegetative system and expressive component
5. Autonomic recruitment	Is important both to prepare the organism to action and in the subjective experience of emotions	Is not requested by a cognitive response
6. Learnig mechanisms	Are based on conditioned automatic and unconscious learning	Are based on the couscious and controlled acquisition of new information in declarative memory

Data reported in Table I are consistent with the general interpretation that Oatley and Johnson-Laird (1987) have given of the main goals of the emotional and of the cognitive system. Thus, with regard to the *analysis of sensory information*, almost all authors acknowledge that the process required to evaluate if an external situation is pleasant or dangerous is usually global, rapid and unconscious. This position is shared not only by biologically oriented authors, such as Ohman (1988), LeDoux (1986, 1992, 1996) and Levenson (1992) but also by supporters of cognitive theories, such as the appraisal model of emotions (Lazarus 1982, 1984; Scherer 1984, 1993, 2000; Frijda 1986, 1987) which maintains that a cognitive evaluation of events is the necessary prerequisite of the ensuing emotion.

Also in agreement with the Oatley and Johnson-Laird's (1987) model is the description of the *action schemata* triggered by the emotionally relevant stimuli that is offered by authors of different orientation, such as Tomkins (1962, 1963, 1984), Ekman (1972, 1973, 1984), Izard (1971, 1990), Leventhal (1979, 1987), Scherer (1984, 1993) and Frijda (1986, 1987). All of these authors acknowledge that emotions must be considered among the most powerful motivational determinants of behaviour and must, therefore, have partly autonomous paths of access to skeletal muscles, to which they transmit patterned drive-related commands. However, in addition to these command mechanisms for the skeletal apparatus, emotions also provoke highly characteristic expressive-motor reactions at the level of the mimic muscles of the face. Data gathered by Tomkins (1962, 1963, 1984), Ekman (1972, 1973, 1984) and Izard (1971, 1990) have shown that these patterned mimic reactions produce distinctive expressions for a limited number of basic emotions (i.e. happiness, sadness, fear, anger, surprise and disgust) and suggest that these expressive reactions are innate. Some authors (e.g. Russell 1980; Larsen and Diener 1992) objected and proposed that, instead of been described in terms of discrete categories, emotions might be more parsimoniously described by only two dimensions, i.e. 'valence' (pleasant vs unpleasant) and 'arousal' (low vs high intensity). However, the same authors (e.g. Diener et al. 1995; Russell and Barrett 1999) subsequently acknowledged that these dimensions are hierarchically related to discrete emotions. We can, therefore, conclude that action schemata, which are innate and activated by an automatic evaluation of an emotionally relevant stimulus, reflect the most important interactive schemata of the human species at the level of communication or of proneness to action. These action schemata tend to solve a certain number of basic problems of the human species in a stereotyped manner, either by activating the most appropriate motor program or by involving the significant person with whom the subject is actually communicating in the solution of the stressing situation. To underline the importance of these social, communicative aspects of emotions, Darwin (1872/1965) rightly pointed out that in man and in other social animals facial and vocal expressions of emotions are innate action patterns, provided with high survival value and widely generalised across the human species.

Another aspect of the emotional response that certainly plays a critical role in emotional behaviour (and can, on the contrary, even disturb the functioning of the cognitive system) is represented by an important recruitment of the vegetative

Box 2.1 Effects of the Sympathetic System at the Level of Different Bodily Organs and Functional Implications of These Effects

Organ	Effect
Eye	Dilation of the pupil, increasing the amplitude of the peripheral visual field
Heart	Increase in rate and force of contraction, thus increase in the availability of oxygen and glucose at the tissue level
Blood vessels	Dilation in brain structures and skeletal muscles selectively involved in emergency responses and constriction of gastrointestinal organs
Lungs	Dilation of bronchioles, increasing the oxygen uptake
Skeletal muscles	Breaking down glycogen stores; immediate release of consumable glucose

nervous system. In spite of the controversies that still exist over the exact role of this autonomic arousal, most authors agree that a state of sympathetic activation acts as a strong determinant of the efficacy of the behavioural response (Cannon 1927; Frijda 1986, 1987; LeDoux 1987, 1996). In particular, Cannon's theory, which mainly accounts for emotions (such as fear or anger) that have strong common adrenergic effects, maintains that the autonomic response serves to prepare the whole organism for action, allowing it to respond quickly and strongly to the emergency situation which has provoked the emotional response. The main effects of sympathetic activity that can influence the efficacy of the behavioural response are reported in Box 2.1.

On the other hand, the classical 'body reaction' theory proposed by James (1884) maintained that the autonomic visceral response, directly provoked by the perception of the emotion eliciting situation, leads to the emotional feelings thanks to a visceral feedback mechanism. Refinements of the 'body reaction' theory, proposed by other authors (e.g. Schachter 1970; Levenson 1992; Damasio 1994; Gu et al 2013), acknowledge that the recruitment of the vegetative nervous system marks an important and specific aspect of the emotional response (the subjective experience of emotions), which is not present (or, in any case is much less relevant) during the execution of tasks involving the cognitive system. Controlled strategic plans, selected by the cognitive system, can, in fact, include intentional (but not an automatic) communicative-expressive components and do not require a concomitant strong activation of the autonomic nervous system.

A final difference between the emotional and the cognitive system concerns the learning mechanisms used by them: Emotional learning is based on unconscious conditioning mechanisms (LeDoux 1992, 1996; Adolphs et al. 1995) whereas the cognitive system makes use of conscious and controlled mechanisms to store new information in declarative memory. All of these data support Oatley and Johnson-Laird's (1987) model, which considers the emotional system as a primitive, but quick and effective, adaptive emergency system.

2.3   The Hierarchical Structure of Human Emotions

In previous sections of this chapter, we have seen that emotions must be considered as a multicomponent hierarchical system. The term 'multicomponent' refers to the fact that the emotional system is based on the integrated activity of sensory, motor (expressive, postural and vegetative), experiential and memory components. The term 'hierarchical' alludes to the fact that different levels of functional organisation have emerged during the phylogenetic development of emotions and that a hierarchical principle is necessary to understand the relationships between these levels. From the phylogenetic point of view, MacLean (1990) has distinguished the most primitive forms of emotional behaviour (such as the fight-flight reactions) which are present in ancient species, such as the reptiles, from the more advanced family related emotional patterns, which are characteristic only of mammals. Particularly important in the context of the present chapter, is the acknowledgement that only the more basic emotions (which appear in humans during the earliest stages of child development) show the characteristics outlined in Table I. By contrast, much more elaborated cognitive evaluations and more flexible and differentiated response options must be taken into account when we refer to complex emotions such as vanity, remorse or shame, which are subtended by social norms and take social rules or the ideal representation of the 'self' as criteria to appraise our behaviour. As most authors (e.g. Ekman 1972, 1984, 1992; Izard 1971, 1990; Leventhal 1987; Montag and Panksepp 2017) sustain that complex emotions derive from basic ones, thanks to mechanisms of blending and of increasing interactions between the emotional and the cognitive systems, acknowledging the difference between elementary and complex emotions has promoted the construction of hierarchically organised psychological models of emotions. A developmental model consistent with this logic and able to explain these facts was formulated by Leventhal (1979, 1987), who hypothesised that human emotions might derive from the activity of a hierarchical multicomponent system based on the activity of three functional levels (the sensorimotor, the schematic and the conceptual level). The main characteristics of this model are reported in Table 2.2.

According to this model, from birth each person disposes of a set of expressive-motor programs that form the first, sensorimotor level of emotions and reflect the basic interactive schemata of the human species at the level of action or of interpersonal communication. During individual development, thanks to mechanisms of conditioned learning, these basic emotional programs are linked to situations of the individual's experience and give rise to the emotional schemata that are the units of the second, schematic level of emotional processing. These schemata differ from the expressive-motor programs of the sensorimotor level, because they are different for each subject, being based on an association between innate and universal programs and events of the individual experience. For instance, a child whose family likes dogs can associate the sight of these animals with an experience of excited amusement. On the contrary, another child whose family does not like pets, can associate the sight of the same animal with an experience of fear. In this manner the same event can give rise to two different emotional schemata in different individuals. Since the automatic

Table 2.2   Main characteristics of the three levels of emotional processing included in Leventhal's developmental model

The Sensori-Motor Level consists of a set of innate expressive-motor programs, which are automatically triggered by a certain number of environmental stimuli and which include components of motor and vegetative activation
The Schematic Level is based on the activity of emotional schemata, i.e. of prototypes of emotional behaviour, formed (on the basis of conditioning processes) by the association between the innate neuromotor programs and situations linked to these programs in individual experience. The afferent parts of the schemata (i.e. the emotion triggering situations) are, therefore, learned and are individual rather than innate and universal. The automatic reactivation of a given emotional schema is accompanied by evocation of the corresponding subjective and expressive-motor components and is experienced as a true emotion
The Conceptual Level is based on a mechanism of conscious learning in declarative memory and is mediated by cognitive processes rather than by conditioning mechanisms. This level stores abstract and propositional notions about the nature of emotions and about the social rules concerning their expression. The activation of these propositional representations is not accompanied by the experience of the corresponding emotion (as is observed during activation of the emotional schemata). The intentional activation made at this level of an 'emotional' expressive-motor program is, therefore, not felt as a true emotion, due to the absence of the corresponding subjective emotional experience.

activation of an emotional schema is accompanied by the evocation of the corresponding emotional experience, the schematic level corresponds to the level of the true, spontaneous, felt emotions and differs, therefore, strikingly from the third, *conceptual level* of emotional processing. The conceptual level is, indeed, based on mechanisms of conscious declarative memory and does not store the concrete representation of true emotions, but rather the abstract and propositional notion of what emotions are, of which situations provoke them and of how to deal appropriately with them, according to the social 'display rules', which establish who can show what emotion to whom and when they can show it (Ekman 1984).

In more recent years, one aspect of the conceptual level of emotional processing that has received increasing attention is the construct of emotion regulation (Gross 1998, 2002; Thayer and Lane 2000; Kim and Hamann 2012; Bechara 2004); it refers to the processes by which we implement (more or less consciously) the goal of starting, stopping or otherwise modulating the trajectory of an emotion (Gross 2015). According to Gross (1998, 2002, 2015), cognitive and behavioural strategies take effect at numerous places along the temporal sequence of emotion generation and these different regulation strategies have different consequences. For instance, reappraisal is a cognitive strategy that has an impact early in the emotion-generative process and alters the trajectory of the emotional response by reformulating the meaning of a situation. Reappraisal decreases emotional experience and behavioural expression but does not affect memory. Another commonly used strategy for down-regulating emotion is suppression, a behavioural strategy which consists of inhibiting the outward signs of inner feelings and is thought to occur even earlier in the emotion generation process. Although decreasing the behavioural expression of emotions, suppression fails to decrease their experience, and actually impairs their memory (Dillon et al. 2007; Katsumi and Dolcos 2018).

References

Adolphs R, Tranel D, Damasio H, Damasio A. Fear and the human amygdala. J Neurosci. 1995;15:5879–81.

Bechara A. The role of emotion in decision-making: evidence from neurological patients with orbitofrontal damage. Brain Cogn. 2004;55:30–40.

Cannon WB. The James-Lange theory of emotion: a critical examination and an alternative theory. Am J Psychol. 1927;39:106–24.

Damasio AR. Descartes' error: emotion, research and the human brain. New York: Avon; 1994.

Darwin C. The expression of the emotions in man and animals. London: Murray; 1872. (reprinted Chicago: University of Chicago Press, 1965)

Diener E, Smith H, Fujita F. The personality structure of affect. J Pers Soc Psychol. 1995;69:130–41.

Dillon DG, Ritchey M, Johnson BD, LaBar KS. Dissociable effects of conscious emotion regulation strategies on explicit and implicit memory. Emotion. 2007;7:354–65.

Ekman P. Universal and cultural differences in facial expression of emotion. In: Cole JR, editor. Nebraska symposium on motivation. Lincoln: University of Nebraska Press; 1972. p. 207–83.

Ekman P. Darwin and cross-cultural studies of facial expression. In: Ekman P, editor. Darwin and facial expression. New York: Academic Press; 1973. p. 1–83.

Ekman P. Expression and the nature of emotion. In: Scherer K, Ekman P, editors. Approaches to emotion. Hillsdale, NJ: Erlbaum; 1984. p. 319–44.

Ekman P. Are there basic emotions? Psychol Rev. 1992;99:550–3.

Frijda NH. The emotions. Cambridge: Cambridge University Press; 1986.

Frijda NH. Emotions, cognitive structures and action tendency. Cogn Emot. 1987;1:115–43.

Gainotti G. Features of emotional behavior relevant to neurobiology and theories of emotions. In: Gainotti G, Caltagirone C, editors. Emotions and the dual brain. Heidelberg: Springer; 1989. p. 9–27.

Gainotti G. Neuropsychological theories of emotions. In: Borod J, editor. The neuropsychology of emotions. New York: Oxford University Press; 2000. p. 147–67.

Gross JJ. The emerging field of emotion regulation: an integrative review. Rev Gen Psychol. 1998;2:271–99.

Gross JJ. Emotion regulation: affective, cognitive, and social consequences. Psychophysiology. 2002;39:281–91.

Gross JJ. Emotion regulation: current status and future prospects. Psychol Inq. 2015;26:1–26.

Gu X, Hof PR, Friston KJ, Fan J. Anterior insular cortex and emotional awareness. J Comp Neurol. 2013;521:3371–88.

Izard CE. The face of emotion. New York: Appleton-Century-Crofts; 1971.

Izard CE. Facial expression and the regulation of emotions. J Pers Soc Psychol. 1990;58:487–98.

James W. What is an emotion ? Mind. 1884;9:188–205.

Katsumi Y, Dolcos S. Suppress to feel and remember less: neural correlates of explicit and implicit emotional suppression on perception and memory. Neuropsychologia. 2018; https://doi.org/10.1016/j.neuropsychologia.2018.02.010.

Kim SH, Hamann S. The effect of cognitive reappraisal on physiological reactivity and emotional memory. Int J Psychophysiol. 2012;83:348–56.

Larsen RJ, Diener E. Promise and problems with the circumplex model of emotions. In: Clark MS, editor. Review of personality and social psychology: emotion, vol. 13. Newbury Park: Sage; 1992. p. 25–9.

Lazarus RS. Thoughts on relations between emotion and cognition. Am Psychol. 1982;37:1019–24.

Lazarus RS. On the primacy of cognition. Am Psychol. 1984;39:124–9.

LeDoux JE. Cognitive-emotional interactions in the brain. Cogn Emotion. 1986;3:267–89.

LeDoux JE. Emotion. In: Mountcastle VB, Plum R, Geiser ST, editors. Handbook of physiology. Sect. I the nervous system, Vol V, higher functions of the brain, part I. Bethesda: American Physiological Society; 1987. p. 419–59.

LeDoux J. Brain mechanisms of emotion and emotional learning. Curr Opin Neurobiol. 1992;2:191–7.

LeDoux J. The emotional brain. New York: Simon and Schuster; 1996.

Levenson RW. Autonomic nervous system differences among emotion. Psychol Sci. 1992;3:23–7.

Leventhal H. A perceptual- motor processing model of emotion. In: Pliner P, Blankestein K, Spiegel IM, editors. Perception of emotion in self and others, vol. 5. New York: Plenum; 1979, p. 1–46.

Leventhal H. A perceptual motor theory of emotion. In: Berkowitz L, editor. Advances in experimental social psychology, vol. 17. New York: Academic Press; 1987. p. 117–82.

MacLean PD. The triune brain in evolution: role in Paleocerebral functions. New York: Plenum; 1990.

Montag C, Panksepp J. Primary emotional systems and personality: an evolutionary perspective. Front Psychol. 2017;8:464. https://doi.org/10.3389/fpsyg.2017.00464.

Oatley K, Johnson-Laird P. Toward a cognitive theory of emotions. Cogn Emot. 1987;1:29–50.

Oatley K, Johnson-Laird P. Cognitive approaches to emotions. Trends Cogn Sci. 2014;18:134–40.

Ohman A. Preattentive processes in the generation of emotion. In: Hamilton V, Bower GH, Frijda NH, editors. Cognitive perspectives on emotion and motivation. Dordrecht: Kluwer Academic Publishers; 1988. p. 127–44.

Plutchik R. Emotion: a psychobioevolutionary synthesis. New York: Harper and Row; 1980.

Russell JA. A circumplex model of affect. J Pers Soc Psychol. 1980;39:1161–8.

Russell JA, Barrett LF. Core affect, prototypical emotional episodes, and other things called emotion: dissecting the elephant. J Pers Soc Psychol. 1999;76:805–19.

Schachter S. The assumption of identity and peripheralist-centralist controversies in motivation and emotion. In: Arnold MB, editor. Feelings and emotions: the Loyola symposium. New York: Academic Press; 1970. p. 111–21.

Scherer KR. On the nature and function of emotion. A component preocess approach. In: Scherer KR, Ekman P, editors. Approaches to emotion. Hillsdale: Erlbaum; 1984. p. 293–318.

Scherer KR. Neuroscience projections to current debates in emotion psychology. Cogn Emot. 1993;7:1–41.

Scherer KR. Psychological models of emotion. In: Borod JC, editor. The neuropsychology of emotion. New York: Oxford University Press; 2000. p. 137–62.

Thayer JF, Lane RD. A model of neurovisceral integration in emotion regulation and dysregulation. J Affect Disord. 2000;61:201–6.

Tomkins SS. Affect, imagery, consciousness: vol. 1, the positive affects. New York: Springer; 1962.

Tomkins SS. Affect, imagery, consciousness: vol. 2, the negative affects. New York: Springer; 1963.

Tomkins SS. Affect theory. In: Scherer KR, Ekman P, editors. Approaches to emotion. Hillsdale, NJ: Erlbaum; 1984. p. 163–96.

Brain Structures Playing a Critical Role in Different Components and Hierarchical Levels of Emotions

3

Contents

3.1 Brain Structures That Underlie the Main Components of Emotions

3.1.1 Brain Structures That Subsume the Evaluation of Emotional Significance

Several authors (e.g. LeDoux 1986, 1992, 1996; LeDoux et al. 1986; Adolphs et al. 1995) have proposed that *the amygdala* could be the structure where conditioned associations are formed between information coming from the external world and internal innate expressive-motor programs and where external stimuli are evaluated in terms of their emotional significance.

Many authors have suggested that the amygdala response to different categories of emotional stimuli may be at least in part specific but the exact features of this specificity are still being investigated. In fact, some authors (e.g. Adolphs et al. 1994, 1995; Calder et al. 1996; Morris et al. 1996; Broks et al. 1998) claimed that the amygdala is specifically involved in recognising fear. Other authors

© Springer Nature Switzerland AG 2020
G. Gainotti, *Emotions and the Right Side of the Brain*,
https://doi.org/10.1007/978-3-030-34090-2_3

(e.g. Paradiso et al. 1999; Anderson et al. 2000) maintained that it contributes to the evaluation of all negative facial expressions. Still other authors surmised that the amygdala is just as important for processing positive reward and reinforcement as it is for negative emotions (e.g. Gottfried et al. 2003; Murray 2007), that the amygdala has a key role in the processing of affective relevance (Ewbank et al. 2009; Murray et al. 2014) or that it detects the intensity of emotions (Bonnet et al. 2015). Consistent with these views are studies (e.g. Adolphs et al. 1999; Siebert et al. 2003) which suggest that the amygdala might be involved in processing biologically relevant stimuli, independently of their valence or category. In any case, some degree of specificity, with greater involvement in recognising fear and, more in general, negative emotions is acknowledged by most authors (e.g. Adolphs et al. 1999; Paradiso et al. 1999; Anderson et al. 2000).

Another interesting view of the functioning of the amygdala derives from a suggestion originally advanced by Papez (1937). Based on this, several authors (i.e. LeDoux 1986, 1996; Gainotti 2001) proposed the existence of two different routes through which emotional stimuli might reach the amygdala. The first (subcortical) route might directly connect the thalamus with the amygdala, transmitting the crude sensory data that are needed to make a quick and raw computation of the possible personal meaning of incoming information. In turn, through its feedback connections with the cortical sensory areas, the amygdala might influence the further processing of incoming information through the second, more complex (cortical) route. A schematic representation of this 'dual route' model is reported in Fig. 3.1.

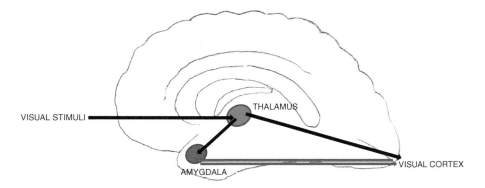

Fig. 3.1 Schematic representation of the 'dual route' model. According to this model, the relay nuclei of the thalamus not only send information to neocortical sensory areas, which forward emotional information to the amygdala (cortical route), but also send fibres directly to the amygdala (subcortical route). The subcortical route transmits to the amygdala the raw sensory data that are needed to make a quick and raw computation of the possible personal meaning of this information. In turn, through its feedback connections with the cortical sensory areas the amygdala can influence the further processing of incoming information through the second, more complex (cortical) route

LeDoux et al. (1986), LeDoux (1996) also showed that the subcortical route has a critical role in processes of emotional conditioning that allow the transition from the 'sensori-motor' to the 'schematic' level of Leventhal's (1979, 1987) model. A lesion of the relay thalamic nuclei disrupts this form of conditioning, whereas ablation of the corresponding cortical sensory areas does not have the same effect.

3.1.2   The Contribution of the Anterior Insula to the Conscious Experience of Emotion

The conscious experience of emotions depends on the perception of bodily reactions to emotion provoking objects and on the cognitive interpretation of these interoceptive informations. The *insular cortex*, which receives in its posterior parts interoceptive inputs from the whole body, and has bidirectional connections in its anterior parts with the frontal, parietal and temporal lobes, the anterior cingulate cortex and the amygdala (Deen et al. 2011) plays an important role in these functions, because its connectivity pattern allows to integrate lower level homeostatic representations with higher level cognitive functions.

Craig (2002, 2005, 2009, 2010, 2011) proposed a posterior-to-anterior gradient in the insular cortex, in which physical features of interoception could be processed in the posterior insula and the integration of interoception with cognitive and motivational information could be made in the anterior insular cortex (AIC). A broad range of somatotopically organised interoceptive information should, therefore, be received by the posterior insula and this interoceptive information should be integrated in the midinsula with input from higher sensory cortices and the limbic system, adding emotional salience to the processing of the interoceptive representation. In the AIC, this representation could be further enriched by integration with input from a large number of cortical areas, forming a unified coherent representation that might be the neural basis for the conscious experience of emotions.

A schematic representation of the integration of homeostatic and cognitive aspects of the conscious experience of emotions in the AIC is reported in Fig. 3.2.

3.1.3   Brain Structures That Contribute to the Generation of Emotional Responses

We have seen in Sect. 2.1. of Chap. 2 that the response programs triggered by emotionally laden stimuli include components of expressive-motor and of autonomic activation, but these two components of the emotional response are probably subtended by partly different brain structures. The autonomic components might be generated by hypothalamic structures, under the control of cortical structures, such as the insula (IC) and the anterior cingulate cortex (ACC), whereas the expressive-motor aspects might be subtended by the ventral striatum (VS) and the ACC.

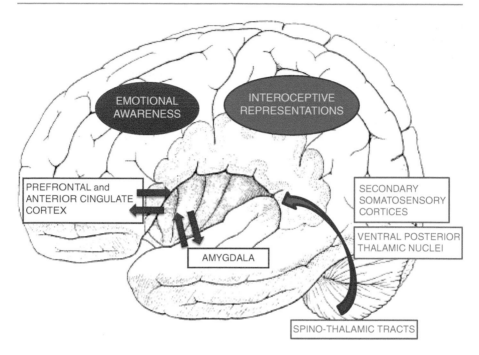

Fig. 3.2 Schematic representation of afferent and efferent connections of the insular cortex and of the integration in its anterior part of homeostatic and cognitive aspects of the conscious experience of emotions. The posterior insula receives interoceptive inputs from the whole body through the spino-thalamic tracts, the ventral posterior thalamic nuclei and the secondary somatosensory cortices. Physical features of interoception are, therefore, processed in the posterior insula, providing first-order homeostatic representations. On the other hand, the anterior insula has bidirectional connections with the frontal, parietal and temporal lobes, the anterior cingulate cortex and the amygdala. These connections allow the integration of lower level homeostatic representations with higher level cognitive functions, forming unified coherent representations that might be the neural basis for the conscious experience of emotions

3.1.3.1 Brain Structures Involved in the Vegetative Components of the Emotional Response

The *hypothalamus* is a brain structure whose major involvement in the generation of vegetative reactions has been known since the Karplus and Kreidl (1909, 1927) experiments. They showed that the electrical stimulation of the hypothalamus is followed by a strong activation of the ortho-sympathetic vegetative system, which is greatly involved in aggression/fear behaviour. Bard (1928) integrated these data with results of midbrain transection experiments and hypothesised that the hypothalamus might have a critical role in emotional behaviour. This proposal was considered as valid for emotions in general and it oriented the attention of authors toward the arousal dimension of emotion. Today the major involvement of the hypothalamus in the generation of the autonomic reaction (reviewed by Smith and DeVito 1984 and more recently by LeDoux 1986, 1996) is universally acknowledged. The regulatory role of the IC and the ACC over the hypothalamic generation of the autonomic response is supported by neuroanatomical, experimental and clinical data. From the neuroanatomical point of view, *the insular cortex* (which integrates nociceptive and visceral inputs) receives afferents from several major autonomic regions, has a viscerotopic sensory organisation and sends efferents to the lateral hypothalamus (Smith and DeVito 1984; Cecchetto and Saper 1990). Its electrical stimulation produces changes in various autonomic parameters both in animals and in humans (Oppenheimer et al. 1992) and its roles in autonomic regulation might be lateralised with a right-sided dominance for sympathetic activities (Oppenheimer et al. 1992; Oppenheimer 1993).

The *anterior cingulate cortex* (which is strongly interconnected with the AIC) is involved in cognitive, motor and autonomic activities and seems necessary to adapt the autonomic state of arousal to concurrent cognitive and physical demands. Devinsky et al. (1995), who distinguished an 'affect' from a 'cognition' component of the ACC, maintained that the affect component has extensive connections with the amygdala and periaqueductal grey, and that parts of it project to autonomic brainstem motor nuclei. The ventral, subgenual parts of the ACC have strong reciprocal connections with the ventral striatal, fronto-orbital and medial temporal regions (Bush et al. 2000) and the genual cingulate, which receives inputs from 'motivational' regions, including the orbitofrontal cortex and amygdala, strongly projects to the hypothalamus (Ongur et al. 1998). Neuroimaging studies have shown that the ACC is part of a network that is systematically activated in a number of disorders that involve emotional dysfunction (e.g. Bush et al. 2000) and during emotional pain (Rainville 2002).

The amygdala is also directly connected to the hypothalamus. Its central nucleus projects mainly to the lateral hypothalamic area (Price and Amaral 1981) and has been implicated in the amygdaloid influence on the autonomic system (Schwaber et al. 1982). A careful description of the direct links between the amygdala and the hypothalamus was provided by Petrovich et al. (2001). These authors assumed that the amygdala modulates instinctive behaviour expression, controlled by the medial hypothalamus. Furthermore, the functional relevance of these links between the amygdala and the hypothalamus (and more in general the impact of the amygdala on autonomic functions) was recently stressed in a study in which Inman et al. (2018) used direct electrical stimulation of the amygdala in humans to examine stimulation-elicited physiological and emotional response. These authors found that in most patients stimulation of the amygdala strongly modulated autonomic activity without necessarily eliciting concurrent subjective emotional responses. Figure 3.3 reports a schematic representation of the connections between hypothalamus, ventral parts of the anterior cingulate cortex and amygdala.

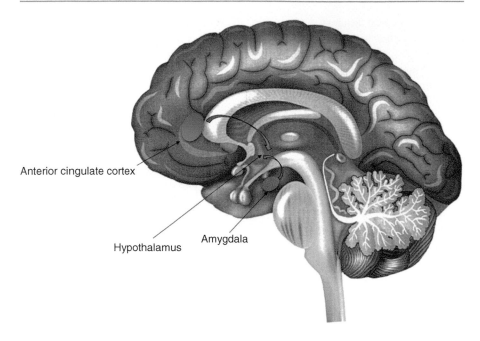

Fig. 3.3 Schematic representation of some descending pathways from the forebrain to the hypothalamus. Besides the insular cortex (that is not reported in this schema), the anterior cingulate cortex and the amygdala are directly connected with the hypothalamus and are involved in modulating the autonomic state in response to the cognitive and physical demands arising from the external milieu

3.1.3.2 Brain Structures Involved Generating the Expressive-Motor Components of the Emotional Response

Cortical and subcortical structures, i.e. the ACC and the VS, are involved in generating the expressive-motor aspects of the emotional response. As I have already surveyed the neuroanatomical and neuroimaging data which suggest including the ACC within the network of the emotion-related structures, here I will discuss only the clinical data that point to a major role of the ACC in the motor components of the emotional response. On one hand, these data include the classical description by Nielsen and Jacobs (1951) and Barris and Schuman (1953) of akinetic mutism with apathy and absence of spontaneous behaviour after ACC lesions; on the other hand, they include more recent observations (Cohen et al. 1994; Szczepanski and Knight 2014) that circumscribed surgical lesions of the ACC are associated with reduced self-generated and spontaneous motor responses.

Involvement of *the ventral striatum* (and of other parts of the basal ganglia) in the execution of stereotyped emotional action patterns is also suggested by anatomical and clinical data. Anatomically, these structures receive strong afferents from the amygdala, the ACC and other limbic structures and send projections to cortical, subcortical and brainstem components of the motor system (Nauta and Domesick

1984). They are strongly connected with orbitofrontal regions (Tekin and Cummings 2002) and amygdala (Cardinal et al. 2002). In particular, it has been suggested that *the nucleus accumbens,* which receives information from a considerable array of limbic structures (including the amygdala) and projects to structures involved in behavioural expression, could represent a 'limbic–motor interface'(Mogenson et al. 1980). According to Cardinal et al. (2002), the nucleus accumbens could have a role in modulating unconditioned behaviours, such as feeding and locomotion, and learned behaviour, including instrumental conditioning, and appears to support animals' ability to work for delayed rewards. Consistent with this account of the brain structures involved in the generation of the expressive-motor aspects of the emotional response are results of a recent review by Le Heron et al. (2018) of the neuroimaging findings in patients with the clinical phenotype of apathy. This condition, conceptualised by Levy and Dubois (2006) as a motivational impairment in goal directed behaviour, was indeed strongly associated with disruption of the ACC, ventral striatum and connected brain structures.

From the clinical point of view, patients with a degenerative disease of the basal ganglia, such as Parkinson's disease, show a marked reduction of spontaneous facial emotional expression (Smith et al. 1996) and analogous defects have been reported in stroke patients with lesions involving the basal ganglia (Cancelliere and Kertesz 1990; Cohen et al. 1994).

Among the brainstem structures which receive afferent connections from the ventral striatum, amygdala and hypothalamus, the most strongly involved in the expression of emotions is probably *the periaqueductal gray.* This rostral brainstem structure sends descending projections to the ventrolateral medulla and to motor nuclei of the cranial nerves and of the spinal cord, which allow to integrate autonomic, expressive-motor and skeleto-motor components of the emotional response in patterns related to general motivational states (Blander and Shipley 1994; Blander and Keay 1996). Figure 3.4 is a schematic illustration of the descending connections to the periaqueductal gray (PAG) from the ventral striatum, hypothalamus and amygdala and of the connections of the PAG with motor nuclei of the cranial nerves and spinal cord structures:

3.2   Brain Structures Involved in the Highest Levels of Emotions and in the Control of Socially Unacceptable Emotional Responses

In Sect. 2.2 of the second chapter we saw that emotions must be considered as a hierarchical system for two main reasons. The first reason is that the phylogenetically more primitive 'basic emotions' must be distinguished from the more evolved 'complex', 'social' and 'self-conscious emotions'. The former are innate and correspond to the Oatley and Johnson-Laird's (1987) definition of an emergency system, based on a small number of modules that automatically process a restricted number of signals and trigger an immediate response. The latter are subtended by social norms and are much more intertwined with (and dependent on) the cognitive system, thus requiring an appreciation of the self in a social context (Tangney 1999).

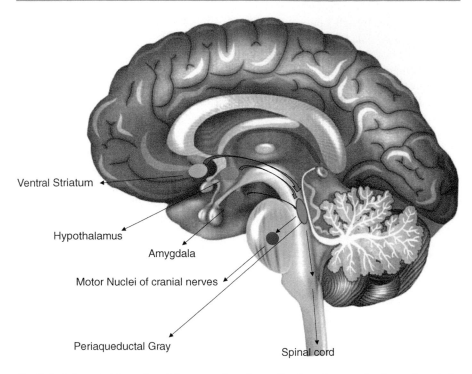

Ventral Striatum

Hypothalamus

Amygdala

Motor Nuclei of cranial nerves

Periaqueductal Gray

Spinal cord

Fig. 3.4 Schematic illustration of the connections descending from the ventral striatum, hypothalamus and amygdala to the periaqueductal gray (PAG) and from the PAG to spinal cord structures and motor nuclei of the cranial nerves. This rostral brainstem structure sends descending projections to the ventrolateral medulla and to the motor nuclei of the cranial nerves; this allows the integration of autonomic, expressive-motor and skeleto-motor components of the emotional response

The second reason is that during ontogenetic development the emotional system becomes more and more interconnected with the cognitive system. As the latter makes use of social 'display rules' (Ekman 1984) to select the most appropriate response to an emotionally laden situation, the action schemata, which are automatically activated by external events, often contrast with these social rules. To solve these conflicts, some cortical structures, lying at the interface between cognitive and emotional system, exert a control over the overt expression of emotions, inhibiting the socially unacceptable emotional outbursts. Since the classical case of patient Pineas Gage, described by Harlow (1868) 150 years ago, many clinical and experimental data have convincingly shown that *the orbitofrontal* (OF) areas and, in particular, their medial regions play a critical role in in the integration between cognition and emotion and in the control of impulsive reactions. Many patients with traumatic or degenerative lesions of these areas have, indeed, shown a severely impaired ability to function in society, even if they obtain normal profiles on standard neuropsychological measures, and are not impaired on cognitive tasks sensitive to frontal lobe damage. (see LeDoux 1996; Bechara et al. 1997, 2000; Drevets and Raichle 1998 and Gainotti 2001 for reviews). Damasio (1994) considered the inability of these patients to integrate their emotional responses into an appropriate social and cognitive context as resulting from a defect in the activation of 'somatic markers'. He used this term (1994) to denote the autonomic and proprioceptive afferences associated with the emotional consequences of a previous personal decision, which converge in the ventromedial prefrontal cortex (vmPFC) and allow us to anticipate the future repercussions of our present actions. A lesion in these cortical areas could disconnect the external stimuli from the corresponding internal 'somatic markers', leaving patients insensitive to the consequences of their abnormal social behaviour. Results supporting these views have been reported by Bechara et al. (1997, 2000). These authors used a gambling task in which choices that yield high immediate gains are followed by higher future losses.

Other interpretations of the role of the OF cortex in emotional behaviour have been offered by authors who have stressed their relevance in processing the more evolved 'social' and 'self-conscious emotions'. According to these views, the OF cortex could be implicated in emotional evaluations that lead to modifying associations previously established at the level of the amygdala, such as fear conditioning (Sotres-Bayon et al. 2006 for review), and in tracking the current reward value for the action of a given stimulus (see Rolls 2004, 2017 for reviews). Thus, while amygdala neurons identify cues with emotional significance in the surrounding world, OF neurons could update the meaning of these cues in current information about the environment and the state of the organism.

In a similar vein, studies of emotional control via cognitive reappraisal, i.e. through the ability to change one's view about a particular situation in order to see it in a different light (see Sect. 2.2 of Chap. 2), have shown that these voluntary aspects of emotion regulation are accomplished through dorsolateral and ventrolateral prefrontal modulation of limbic-mediated emotional reactivity (e.g. Fitzgerald et al. 2018). This contrasts with observations made during the expressive suppression condition, in which decreased negative emotional behaviour and experience are accompanied by increased sympathetic activation and by enhanced amygdala and insula responses (Goldin et al. 2008). These data are consistent with results of neuroimaging studies (surveyed by Buhle et al. 2014) which have shown that the amygdala is modulated by areas of the prefrontal cortex supporting cognitive reappraisal of aversive stimuli.

3.3   More Complex Neurobiological Models of the Interactions Between Different Brain Structures Involved in Emotional Functions

In previous sections of this chapter, I have presented a simplified and schematic illustration of the dissimilar and complementary roles that distinct brain structures play in different components and hierarchical levels of emotions. At the end of the last section I have, however, acknowledged that complex interactions exist between these structures, reporting the case of the modulatory activity that areas of the prefrontal cortex exert on the evaluation of emotional information accomplished by amygdala. In the present section I intend to dwell on a different aspect of this important problem, namely the role that, according to Adolphs et al. (2000), right-sided somatosensory and insular cortices could play in recognition of visual emotional stimuli. Drawing on previous studies (e.g. Bowers et al. 1985; Adolphs et al. 1994; Gur et al. 1994; Borod et al. 1998) which had stressed the importance of right hemisphere cortices in the recognition of facial emotion, these authors investigated the association between lesion location and facial emotion recognition in 108 patients with focal brain damage and provided evidence that somatosensory cortices and insula in the right hemisphere play a critical role in facial emotion recognition. To explain these findings, the authors assumed, in agreement with Wild et al. (2001) that viewing facial expressions of emotion may trigger an emotional response in the perceiver that mirrors the emotion shown by the stimulus. The induction of this emotional response might, in turn lead to somatic and vegetative changes, that could be detected by the somatosensory and insular cortices, providing information about the seen emotion. It is, therefore, possible that the recognition of facial emotion may not necessarily involve only the visual cortex and the amygdala. It could also trigger a complex chain of reactions, including the generation of the same emotion via the process of simulation (Gallese et al. 2004), which could be decoded by insula and somatosensory cortices, allowing knowledge about the seen facial emotional expression.

References

Adolphs R, Tranel D, Damasio H, Damasio A. Impaired recognition of emotion in facial expressions following bilateral damage to the human amygdala. Nature. 1994;372:669–72.

Adolphs R, Tranel D, Damasio H, Damasio A. Fear and the human amygdala. J Neurosci. 1995;15:5879–81.

Adolphs R, Tranel D, Hamann S, Young AW, Calder AJ, Phelps EA, et al. Recognition of facial emotion in nine subjects with bilateral amygdala damage. Neuropsychologia. 1999;37:1111–7.

Adolphs R, Damasio H, Tranel D, Cooper G, Damasio AR. A role for somatosensory cortices in the visual recognition of emotion as revealed by three-dimensional lesion mapping. J Neurosci. 2000;20:2683–90.

Anderson AK, Spencer DD, Fulbright RK, Phelps EA. Contribution of the anteromedial temporal lobes to the evaluation of facial emotion. Neuropsychology. 2000;14:526–36.

Bard P. A diencephalic mechanism for the expression of rage with special reference to the sympathetic nervous system. Am J Physiology. 1928;84:490–515.

Barris RW, Schuman HR. Bilateral anterior cingulated gyrus lesions: syndrome of the anterior cingulate gyri. Neurology. 1953;3:44–52.

Bechara A, Damasio H, Tranel D, Damasio AR. Deciding advantageously before knowing the advantageous strategy. Science. 1997;275:1293–5.

Bechara A, Tranel D, Damasio H. Characterization of decision-making deficit of patients with ventromedial prefrontal cortex. Brain. 2000;123:2189–202.

Blander R, Keay KA. Columnar organization in the midbrain periaqueductal gray and the integration of emotional expression. In: Holstege G, Bandler R, Saper CB, editors. Progress in brain research, The emotional motor system, vol. 107. Amsterdam: Elsevier; 1996. p. 285–300.

Blander R, Shipley MT. Columnar organization in the midbrain periaqueductal gray: modules for emotional expression ? Trends Neurosci. 1994;17:379–89.

Bonnet L, Comte A, Tatu L, Millot JL, Moulin T, Medeiros de Bustos E. The role of the amygdala in the perception of positive emotions: an "intensity detector". Front Behav Neurosci. 2015;9:178. https://doi.org/10.3389/fnbeh.2015.00178.

Borod JC, Obler LK, Erhan HM, Grunwald IS, Cicero BA, Welkowitz J, et al. Right hemisphere emotional perception: evidence across multiple channels. Neuropsychology. 1998;12:446–58.

Bowers D, Bauer RM, Coslett HB, Heilman KM. Processing of faces by patients with unilateral hemisphere lesions. Brain Cogn. 1985;4:258–72.

Broks P, Young AW, Maratos EJ, Coffey PJ, Calder AJ, Isaac CL, et al. Face processing impairments after encephalitis: amygdala damage and recognition of fear. Neuropsychologia. 1998;36:59–70.

Buhle JT, Silvers JA, Wager TD, Lopez R, Onyemekwu C, Kober H, et al. Cognitive reappraisal of emotion: a meta-analysis of human neuroimaging studies. Cer Cortex. 2014;24: 2981–90.

Bush G, Luu P, Poster MI. Cognitive and emotional influences in anterior cingolate cortex. Trends Cogn Sci. 2000;4:215–22.

Calder AJ, Young AW, Rowland D, Perret DI. Facial emotion recognition after bilateral amygdala damage: differentially severe impairment of fear. Cogn Neuropsychol. 1996;13:699–745.

Cancelliere AEB, Kertesz A. Lesion localization in acquired deficits of emotional expression and comprehension. Brain Cogn. 1990;13:133–47.

Cardinal RN, Parkinson JA, Hall J, Everitt BJ. Emotion and motivation: the role of the amygdala, ventral striatum, and prefrontal cortex. Neurosci Biobehav Rev. 2002;26:321–52.

Cecchetto DF, Saper CB. Role of the cerebral cortex in autonomic function. In: Loewy AD, Spyer KM, editors. Central regulation of autonomic functions. New York: Oxford University Press; 1990. p. 208–23.

Cohen MJ, Riccio CA, Falnnery AM. Expressive aprosodia following stroke to the right basal ganglia: a case report. Neuropsychology. 1994;8:242–5.

Craig AD. A new view of pain as a homeostatic emotion. Trends Neurosci. 2002;26:303–7.

Craig AD. Forebrain emotional asymmetry: a neuroanatomical basis? Trends Cogn Sci. 2005;9:566–71.

Craig AD. How do you feel—now? The anterior insula and human awareness. Nat Rev Neurosci. 2009;10:59–70.

Craig AD. The sentient self. Brain Struct Funct. 2010;214:563–77.

Craig AD. Significance of the insula for the evolution of human awareness of feelings from the body. Ann N Y Acad Sci. 2011;1225:72–82.

Damasio AR. Descartes' error: emotion, research and the human brain. New York: Avon; 1994.

Deen B, Pitskel NB, Pelphrey KA. Three systems of insular functional connectivity identified with cluster analysis. Cereb Cortex. 2011;21:1498–506.

Devinsky O, Morrel MJ, Vogt BA. Contributions of anterior cingulate cortex to behaviour. Brain. 1995;118:279–306.

Drevets WC, Raichle ME. Reciprocal suppression of regional cerebral blood flow during emotional versus higher cognitive processes: implications for interactions between cognition and emotion. Cogn Emot. 1998;12:353–85.

Ekman P. Expression and the nature of emotion. In: Scherer K, Ekman P, editors. Approachs to emotion. Hillsdale, NJ: Erlbaum; 1984. p. 319–44.

Ewbank MP, Barnard PJ, Croucher CJ, Ramponi C, Calder AJ. The amygdala response to images with impact. Soc Cogn Affective Neurosci. 2009;4:127–33.

Fitzgerald JM, Kinney KL, Phan KL, Klumpp H. Distinct neural engagement during implicit and explicit regulation of negative stimuli. Neuropsychologia. 2018; https://doi.org/10.1016/j.neuropsychologia.2018.02.002.

Gainotti G. Emotions as a biologically adaptive system: an introduction. In: Gainotti G, editor. Handbook of neuropsychology, vol. 5. 2nd ed. Emotional Behaviour and its Disorders. Amsterdam: Elsevier; 2001. p. 1–15.

Gallese V, Keysers C, Rizzolatti G. A unifying view of the basis of social cognition. Trends Cogn Sci. 2004;8:396–403.

Goldin PR, Mcrae K, Ramel W, Gross JJ. The neural bases of emotion regulation: reappraisal and suppression of negative emotion. Biol Psychiatry. 2008;63:577–86.

Gottfried JA, O'Doherty J, Dolan RJ. Encoding predictive reward value in human amygdala and orbitofrontal cortex. Science. 2003;301(5636):1104–7.

Gur RC, Skolnick BE, Gur RE. Effects of emotional discrimination tasks on cerebral blood flow: regional activation and its relation to performance. Brain Cogn. 1994;25:271–86.

Harlow JM. Recovery from the passage of an iron bar through the head. Publ Mass Med Soc. 1868;2:237–46.

Inman CS, Bijanki KR, Bass DI, Gross RE, Hamann S, Willie JT. Human amygdala stimulation effects on emotion physiology and emotional experience. Neuropsychologia. 2018; https://doi.org/10.1016/j.neuropsychologia.2018.03.019.

Karplus JP, Kreidl A. Gehirn und Sympathicus. I Zwischenhirnbasis und Hallsympathicus Pf Arch Gesam Physiol Men Thiere. 1909;129:138–44.

Karplus JP, Kreidl A. Gehirn und Sympathicus. VII Uber Beziehungen der Hypothalamuszentren zu Blutdruck und innerer Sekretion. Pf Arch Gesam Physiol Men Thiere. 1927;215:667–70.

Le Heron C, Apps MAJ, Husain M. The anatomy of apathy: a neurocognitive framework for a motivated behaviour. Neuropsychologia. 2018;118:54–67.

LeDoux JE. Cognitive-emotional interactions in the brain. Cogn Emotion. 1986;3:267–89.

LeDoux J. Brain mechanisms of emotion and emotional learning. Curr Opin Neurobiol. 1992;2:191–7.

LeDoux J. The emotional brain. New York: Simon and Schuster; 1996.

LeDoux JE, Sakaguchi A, Iwata J, Reis DJ. Interruption of projections from the medial geniculate body to an archi-neostriatal field disrupts the classical conditioning of emotional responses to acoustic stimuli in the rat. Neuroscience. 1986;17:615–27.

Leventhal H. A perceptual- motor processing model of emotion. In: Pliner P, Blankestein K, Spiegel IM, editors. Perception of emotion in self and others, vol. 5. New York: Plenum; 1979. p. 1–46.

Leventhal H. A perceptual motor theory of emotion. In: Berkowitz L, editor. Advances in experimental social psychology, vol. 17. New York: Academic Press; 1987. p. 117–82.

Levy R, Dubois B. Apathy and the functional anatomy of the prefrontal cortex-basal ganglia circuits. Cereb Cortex. 2006;16:916. https://doi.org/10.1093/cercor/bhj043.

Mogenson GJ, Jones DL, Yim CY. From motivation to action: functional interface between the limbic system and the motor system. Prog Neurobiol. 1980;14:69–97.

Morris JS, Frith CD, Perrett DI, et al. A differential neural response in the human amygdala to fearful and happy facial expressions. Nature. 1996;383:812–5.

Murray EA. The amygdala, reward and emotion. Trends Cogn Sci. 2007;11:489–97.

Murray R, Brosch T, Sander D. The functional profile of the human amygdala in affective processing: insights from intracranial recordings. Cortex. 2014;60:10–33.

Nauta WJH, Domesick VB. Afferent relationships of the basal ganglia. Ciba Found Symp. 1984;107:3–23.

Nielsen JM, Jacobs LL. Bilateral lesions of the anterior cingulate gyri. Report of case Bull Los Angeles Neurol Soc. 1951;16:231–4.

Oatley K, Johnson-Laird P. Toward a cognitive theory of emotions. Cogn Emot. 1987;1:29–50.

Ongur D, An X, Price JL. Prefrontal cortical projections to the hypothalamus in macaque minkeys. J Comp Neurol. 1998;40:480–505.

Oppenheimer SM. The anatomy and physiology of cortical mechanisms of cardiac control. Stroke. 1993;24:13–5.

Oppenheimer SM, Gelb A, Girvin JP, Hachinski VC. Cardiovascular effects of human insular cortex stimulation. Neurology. 1992;42:1727–32.

Papez JW. A proposed mechanism of emotion. Arch Neurol Psychiatr. 1937;79:217–24.

Paradiso S, Johnson DL, Andreasen NC, O'Leary DS, Watkins GL, Ponto LL. Cerebral blood flow changes associated with attribution of emotional valence to pleasant, unpleasant, and neutral visual stimuli in a PET study of normal subjects. Am J Psychiatry. 1999;156:1618–29.

Petrovich GD, Canteras NS, Swanson LW. Combinatorial amygdalar inputs to hippocampal domains and hypothalamic behavior systems. Brain Res Brain Res Rev. 2001;38:247–89.

Price JL, Amaral DG. An autoradiographic study of the projections of the central nucleus of the monkey amygdala. J Neurosci. 1981;1:1242–59.

Rainville P. Brain mechanisms of pain affect and pain modulation. Curr Opin Neurobiol. 2002;12:195–204.

Rolls ET. The functions of the orbitofrontal cortex. Brain Cogn. 2004;55:11–29.

Rolls ET. The orbitofrontal cortex and emotion in health and disease, including depression. Neuropsychologia. 2017;128:14. https://doi.org/10.1016/j.neuropsychologia.2017.09.021.

Schwaber JS, Kapp BS, Higgins GA, Rapp PR. Amygdaloid and basal forebrain direct connections with the nucleus of the solitary tract and the dorsal motor nucleus. J Neurosci. 1982;2:1424–38.

Siebert M, Markowitsch HJ, Bartel P. Amygdala, affect and cognition: evidence from 10 patients with Urbach-Wiethe disease. Brain. 2003;126:2627–37.

Smith OA, DeVito JL. Central neural integration for the control of autonomic responses associated with emotion. Annu Rev Neurosci. 1984;7:43–65.

Smith MC, Smith MK, Ellgring H. Spontaneous and posed facial expression in Parkinson's disease. J Int Neuropsychol Soc. 1996;2:383–91.

Sotres-Bayon F, Cain CK, LeDoux JE. Brain mechanisms of fear extinction: historical perspectives on the contribution of prefrontal cortex. Biol Psychiatry. 2006;60:329–36.

Szczepanski SM, Knight RT. Insights into human behavior from lesions to the prefrontal cortex. Neuron. 2014;83:1002–18.

Tangney JP. The self-conscious emotions: shame, guilt, embarassment and pride. In: Dalgleish T, Power MJ, editors. Handbook of cognition and emotion. New York: Wiley; 1999.

Tekin S, Cummings JL. Frontal-subcortical neuronal circuits and clinical neuropsychiatry. an update J Psychosom Res. 2002;53:647–54.

Wild B, Erb M, Bartels M. Are emotions contagious? Evoked emotions while viewing emotionally expressive faces: quality, quantity, time course and gender differences. Psychiatry Res. 2001;102:109–24.

The History of Research on Emotional Laterality

4

Contents

A rather recent hypothesis assumes that in addition to the classical distinction between cortical and subcortical structures the distinction between the left and right halves of the brain also has an important role in explaining the representation and control of human emotions. This is particularly impressive if we consider the time that has elapsed between the first studies showing a left hemisphere dominance for

© Springer Nature Switzerland AG 2020   27
G. Gainotti, *Emotions and the Right Side of the Brain*,
https://doi.org/10.1007/978-3-030-34090-2_4

language (Dax 1836, quoted by Dax 1865; Broca 1865) and the first reports of laterality effects in the representation of emotions that had a clear impact on following investigations (Terzian and Cecotto 1958; Alemà and Donini 1960; Perria et al. 1961; Terzian 1964; Rossi and Rosadini 1967; Gainotti 1969, 1972). It must be acknowledged, however, that even before these reports, the different emotional behaviour of patients affected by unilateral brain lesions (and other related clinical phenomena) had repeatedly attracted the attention of skilled clinicians and researchers, because they suggested a right hemisphere dominance for emotions or, in any case, an asymmetrical representation of emotions in the human brain.

In the present chapter, therefore, the discussion of recent investigations of the links between emotions and the right hemisphere will be preceded by an historical section, aimed at placing the results of these studies in an appropriate historical and theoretical context. More precisely, I will review first some important pioneering observations on the relationships between emotions and cerebral dominance. Then I will survey the clinical investigations which have suggested a general dominance of the right hemisphere for all kinds of emotions and the experimental studies which have proposed a different hemispheric specialisation for positive vs negative emotions; finally I will discuss models advanced to explain results of these studies.

4.1   The Pioneers

According to the historical surveys of Harrington (1995) and Harris (1999), the first author who suggested a hemispheric asymmetry in the representation of emotions was Jules Bernard Luys.

In series of papers and books (Luys 1879, 1881, 1893), this author reported differences between the emotional behaviour of right and left brain-damaged patients and proposed the existence of an 'emotion' centre in the right hemisphere, complementary to the 'intellectual' centres in the left hemisphere. However, this proposal went unnoticed and did not influence subsequent investigations (Harris 1999). In fact, more than a century elapsed from the time when the scientific community was interested by the first studies showing that the left hemisphere (LH) is dominant for language (Dax 1836, quoted by Dax 1865; Broca 1865) and the time when the asymmetrical representation of emotions in the human brain was inserted in a context of scientific investigations concerning the brain and its pathology. This occurred after observations on this topic by Terzian and Cecotto (1958), Alemà and Donini (1960), Perria et al. (1961) and Gainotti (1969, 1972). Although Luys' very provocative proposal went substantially unnoticed (Gainotti 2019), other clinical observations relevant to this issue were made by different authors between Luys' (1879, 1881, 1893) publications and Terzian and Cecotto (1958) paper. For example on the basis of his clinical studies of patients with unilateral brain damage, Mills (1912) suggested that deficits in emotional expression might be frequently linked to right hemisphere lesions; and Babinski (1914) observed that unawareness of organic hemiplegia (anosognosia) and indifference toward the resulting disability (anosodiaphoria) are usually due to right hemisphere damage. Furthermore, discussing the relationships between anosognosia and right hemisphere lesions, Hirschl and Potzl

(quoted by Schilder 1935) advanced the hypothesis of a right hemisphere dominance for vegetative functions. However, this idea was not taken into account by
contemporary authors because it was not in line with the prevailing concepts about
cerebral dominance. Some years later, Goldstein (1939) described a very different
kind of emotional reaction to the consequences of brain injury in left brain-damaged
(and particularly in aphasic) patients. These subjects reacted to difficulties of verbal
expression or other difficulties met during the neuropsychological examination with
a pattern of behaviour that Goldstein termed a 'catastrophical reaction'; it consisted
of the inability to control the situation, with increasing signs of anxiety and sometimes with overt outbursts of crying. However, Goldstein did not stress the relationship between catastrophic reactions and left hemispheric lesions and, thus, his
observations were not influential with respect to the problem of emotion lateralisation. The same remark also applies to Critchley's (1955, 1957) observation that
right brain-damaged patients often show an abnormal reaction to the consequences
of their brain lesion. This reaction not only consists of unawareness of the disability
(anosognosia) and of indifference toward it (anosodiaphoria), but also of other paradoxical attitudes toward the hemiplegia. Among these attitudes, Critchley (1955,
1957) described the personification of the paralised limbs and an almost insane
hatred toward the paralysed extremity that was couched in exaggerated language
(misoplegia). Thus, in spite of other anecdotal observations about the abnormal
emotional behaviour of right brain-damaged patients made by different authors,
until the beginning of the 1960s the right and left hemispheres were considered to
be equally involved in the representation of emotions. Reviewing at an international
conference on interhemispheric relations and cerebral dominance the clinical symptomatology of patients with right and left brain lesions, Hécaen (1962: 217) could,
therefore, say that 'when there is a disturbance of the activities of synthesis (such as
personality disturbances derived from cerebral lesions) no difference can be found
between left and right lesions'. At almost the same time, however, Terzian and
Cecotto (1958), Alemà and Donini (1960) and Perria et al. (1961) noticed that the
emotional reactions of patients submitted to pharmacological inactivation of the
right and left hemisphere were different according to the side of the injection. In
fact, a 'depressive-catastrophic' reaction was observed after pharmacological inactivation of the left hemisphere, and an opposite 'euphoric-manic' reaction was produced by injection of sodium amytal into the right carotid artery. These systematic
observations raised the problem of the relationships between emotions and cerebral
dominance in a stable manner.

4.2   First Interpretations of the Different Emotional Behaviour Shown by Right and Left Brain-Damaged Patients

The authors who had observed a different emotional reaction in patients submitted
to right and left intracarotid amytal injection interpreted their data making implicit
reference to the biological model of the bipolar manic-depressive psychosis. They
reported that 'depressive-catastrophic' reactions observed after injection of sodium

amytal into the left carotid artery were characterised by a sad and pessimistic attitude, with tendency to cry, whereas the 'euphoric-manic' reactions observed after
inactivation of the right hemisphere were characterised by relaxed behaviour, with
a tendency to joke and to break into laughter (Terzian and Cecotto 1958; Alemà
and Donini 1960; Perria et al. 1961; Terzian 1964; Rossi and Rosadini 1967).
These different emotional reactions were interpreted as due to the inactivation of
two neural mechanisms that underpin opposite aspects of mood and that are located
respectively in the right and left hemispheres. According to this model: (a) normal
mood should result from a balance between a left-sided centre for positive and a
right-sided centre for negative emotions, (b) depressive-catastrophic reactions
observed after inactivation of the left hemisphere should reflect the prevalence of
the right-sided centre for negative emotions, and (c) euphoric-manic reactions
observed after injection of sodium amytal into the right carotid artery should be
due to the dominance of the left-sided centre for positive emotions. Some years
later, Gainotti (1969, 1972) was only able to partially confirm observations made
during pharmacological inactivation of the right and left hemisphere, by analysing
the patterns of emotional behaviour shown by right and left brain-damaged patients
during a neuropsychological examination. A 'catastrophic reaction' was, indeed,
typically observed in (aphasic) left brain-damaged patients, whereas an 'indifference reaction' was usually found in patients with severe RH lesions. However, the
qualitative features of these reactions and the typical context in which they had
been observed led Gainotti (1969, 1972) to reject the interpretation linking 'catastrophic reactions' to a biological depression and 'indifference reactions' to a
manic state. Catastrophic reactions were, indeed, usually triggered by repeated,
frustrating attempts of verbal expression and looked like an emotional storm, rather
than a depressive orientation of the mood. They were, therefore, considered as a
dramatic, but psychologically appropriate form of reaction to a catastrophic event,
rather than as a biological form of depression. Indifference reactions of right brain-
damaged patients, on the other hand, were not characterised by euphoric mood,
talkativeness and agitation (typical of manic states), but rather consisted of a heterogeneous and rather paradoxical set of emotional abnormalities. A first pattern of
behaviour included overt expressions of denial or minimisation of the disease, a
lack of concern for the disability and an attitude of indifference toward failures met
during the examination. A second behavioural pattern consisted of a tendency to
joke in a fatuous, ironic or sarcastic manner, which sometimes gave the impression
of black humour. A final set of abnormal emotional reactions included delusions
(somatoparaphrenia) and expressions of hatred toward the paralysed limbs (misoplegia), usually couched in a grotesque, exaggerated or sarcastic manner. As all
these behavioural patterns were much more abnormal than the catastrophic reactions of left brain-damaged patients, Gainotti (1972) proposed that the difference
between the emotional behaviour of right and left brain-damaged patients could be
better captured by the contrast between 'appropriate' and 'inappropriate' emotional reactions than by the opposition between 'depressive' and 'euphoric' mood
states. According to this interpretation, the abnormal emotional reaction of right
brain-damaged patients could be due to a major involvement of the right

hemisphere in emotional functions, just as the language disturbances of left brain-damaged patients are due to the left hemisphere dominance for language. When the right hemisphere is intact (as in left brain-damaged patients), the emotional reaction can be dramatic, but is emotionally appropriate. When, on the contrary, the lesion disrupts right hemisphere structures that are critically involved in emotional, then a composite set of abnormal emotional reactions, grouped for simplicity's sake under the heading of 'indifference reactions', is often observed. Thus, the first clinical observations that raised the problem of hemispheric asymmetries in emotion representation also prompted the two models of emotional lateralisation that are still considered as the most important ones in this area of inquiry and that can be labelled as *the different hemispheric specialisation* (or *'the valence'*) *hypothesis* and *the 'right hemisphere dominance' hypothesis.*

4.3   Experimental and Clinical Investigations That Have Studied the Nonverbal Communicative Aspects of Emotions

In the following years many experimental and clinical investigations were conducted in normal subjects and in patients with unilateral brain damage to clarify the meaning of the different emotional reactions reported by Terzian and Cecotto (1958), Alemà and Donini (1960), Perria et al. (1961), Terzian (1964) and Rossi and Rosadini (1967) after pharmacological inactivation of the right and left hemisphere and by Gainotti (1969, 1972) during the neuropsychological examination of right and left brain-damaged patients. Most of these investigations contrasted the *'valence'* with the *'right hemisphere dominance'* hypothesis using methodologies that evaluated the communicative aspects of emotions. There were probably several reasons for the attention given to these aspects of emotions: (a) the rediscovery of Darwin's (1872) seminal work by Tomkins (1962, 1963) and Ekman (1972, 1984) and the studies of these authors showing that cross-cultural patterns of facial expression can be found for a number of basic emotions; (b) the parallel development of sophisticated techniques of analysis of nonverbal communication, such as the FACS procedure, developed by Ekman and Friesen (1978); (c) the tendency to assume that hemispheric asymmetries emerge in the more complex (in this case the communicative) rather than in the more elementary components of a given function. As it is easier to study the positive and negative components of emotions by contrasting their expression at the facial (mimicry) or the vocal (prosodic) level, most of these investigations used methodologies that evaluated comprehension or production of positive and negative emotions at the level of mimicry or emotional prosody. Thus, in research conducted on normal subjects, the mono-hemispheric treatment of emotional information was studied both in the visual modality (lateralised tachistoscopic presentation) and in the auditory modality (dichotic listening or lateralised auditory presentation), whereas the hemispheric asymmetries on the executive side were studied primarily by comparing the emotional expressiveness of the two halves of the face. On the other hand, in research conducted on brain-damaged patients, the influence of unilateral lesions on comprehension and expression of positive and negative emotions was evaluated comparing the ability of these patient groups to recognise or express emotions with both voice (emotional prosody) and facial expressions.

4.3.1 Investigations That Have Studied Comprehension and Expression of Emotions in Normal Subjects

Using the *dichotic listening technique* to evaluate the comprehension of auditory emotional stimuli, Haggard and Parkinson (1971), Carmon and Nachshon (1973), Ley and Bryden (1982) and Bryden (1982) showed an overall left ear (right hemisphere) advantage in the recognition of nonverbal emotional sounds and emotional prosody. Furthermore, Safer and Leventhal (1977) and Zenhausern et al. (1981) showed that if one sends verbal messages in which there is a contrast between semantic content ('what is said') and emotional prosody ('how it is said') to the left and right ears, the right hear (left hemisphere) response will reflects the verbal content, and the left ear (right hemisphere) response will reflect the emotional prosody. Thus if the sentence 'I love you', pronounced with a tone of voice expressing hatred, is presented to both ears and the question is: 'What does he/she feel?', the response to the right ear presentation will be 'he loves her', whereas the response to the left ear presentation will be 'he hates her'.

Consistent with these data were results obtained by Suberi and McKeever (1977), Buchtel et al. (1978), Landis et al. (1979), Ley and Bryden (1979), Hansch and Pirozzolo (1980), Làdavas et al. (1980), Strauss and Moscovitch (1981), Heller and Levy (1981), Duda and Brown (1984), Kinsbourne and Bemporad (1984), Mandal and Singh (1990) and Hugdahl et al. (1993). These authors used a *tachistoscopic lateralised presentation of emotional faces to the right and left visual field* to evaluate the recognition of facial emotional expressions. All of these authors found an overall left visual field (right hemisphere) advantage for emotional faces. Using a similar technique, Reuter-Lorenz and Davidson (1981), Reuter-Lorenz et al. (1983), and Natale et al. (1983) documented, however, a right hemisphere advantage for negative emotions and a left hemisphere advantage for positive emotions. Furthermore, Asthana and Mandal (2001) found a right hemisphere superiority for the recognition sad facial emotion, but not in the perception of happy expression. Contrasting results were also obtained by Wittling and Roscmann (1993) and by Dimond and Farrington (1977) who studied the responsiveness to emotional stimuli with a lateralised presentation of positive and negative films. The former documented a higher subjective emotional response to both kinds of emotional stimuli when they were shown to the right hemisphere, whereas the latter found a greater response of the right hemisphere to a sad film and of the left hemisphere to a funny film. Results similar to those obtained evaluating the recognition of facial emotional expressions were obtained studying facial asymmetries in the expression of emotions. Thus, Moscovitch and Olds (1982), Dopson et al. (1984) and Wylie and Goodale (1988) studied facial expression of emotions with video records of spontaneous emotions and demonstrated a general right hemisphere dominance for facial expression of felt emotions. Similar results were obtained by Campbell (1978), Sackeim and Gur (1978), Sackeim et al. (1978), Heller and Levy (1981), Sackeim and Grega (1987), Borod et al. (1988), Moreno et al. (1990), Asthana and Mandal (1998), and Nicholls et al. (2002) using *chimeric faces* (i.e. composite photographs of posed positive or negative emotions, created using two left or two right half faces). These authors found that the left side of the face (presumably controlled by the right cerebral hemisphere) moves more extensively and appears more intense during emotional expression than the right hemiface. Both positive and negative left-left composites were, therefore, judged to express emotions more intensely than the right-right composites. These results were, however, criticised by Hager and Ekman (1985), who studied the asymmetry of several different muscular actions under different eliciting conditions and claimed that the asymmetry of facial actions is inconsistent with models of hemispheric specialisation.

In fact, they found that during deliberate actions, some of the asymmetries were lateralised with greater intensity on the left side and others, on the right side and that spontaneous actions were more symmetrical than the deliberate, requested actions.

4.3.2   Investigations That Studied the Communicative Aspects of Emotions in Right and Left Brain-Damaged Patients

Many, but not all, studies that investigated the capacity of right and left brain-damaged patients to comprehend emotions expressed with the tone of voice or at the facial level found greater impairment of patients with right-sided lesions. Thus, Heilman et al. (1975), Tucker et al. (1977), Ross (1981), Heilman et al. (1984), Blonder et al. (1991), Schmitt et al. (1997) and Ross and Monnot (2008) showed that right brain-damaged patients understand the linguistic components of a verbal message but cannot appreciate the emotional tone of voice of the speaker. However, Schlanger et al. (1976), Brådvik et al. (1990) and Cancelliere and Kertesz (1990) did not find a significant difference between right and left brain-damaged patients on emotional prosody tasks.

Similarly, DeKosky et al. (1980), Cicone et al. (1980), Benowitz et al. (1983), Etcoff (1984), Bowers et al. (1985), Borod et al. (1986), Blonder et al. (1993), Mandal et al. (1993) and Schmitt et al. (1997) showed that right brain-damaged patients obtain very poor scores on tasks of identification of the emotion displayed by a given face. However, Schlanger et al. (1976), Prigatano and Pribram (1982), Weddell (1989) and Peper and Irle (1997) were unable to find significant differences between patients with right and left brain lesions on this kind of task (see Gainotti 1989, Gainotti et al. 1993, and Schmitt et al. 1997 for reviews). A very similar picture emerges if we pass from the receptive to the expressive level of emotional communication. Tucker et al. (1977), Ross and Mesulam (1979), Ross (1981, 1984), Hughes et al. (1983), Borod et al. (1985), Gorelick and Ross (1987), Ross et al. (1997) and Ross and Monnot (2008) showed that right brain-damaged patients are unable to *express their feelings with the emotional tone of voice*, but Brådvik et al.

(1990) and Cancelliere and Kertesz (1990) failed to find a significant difference between right and left brain-damaged patients on this subject.

Analogously, in studies of spontaneous facial expression, Ross and Mesulam (1979), Buck and Duffy (1980) Borod et al. (1985, 1988) and Blonder et al. (1993) showed that right brain-damaged patients are poor at expressing emotions with gestures or facial expression, but Kolb and Milner (1981), Mammucari et al. (1988) and Weddell (1989) were unable to find significant differences on this subject between patients with right and left hemispheric lesions.

Finally, three studies (Bruyer 1981; Borod et al. 1986, 1990) that examined posed facial expressions of emotions found greater impairment in patients with right brain damage, but three other studies (Caltagirone et al. 1989a; Kolb and Taylor 1981; Weddell et al. 1990) found no difference on this subject between right and left brain-damaged patients.

Some of these studies (e.g. Buck and Duffy 1980; Borod et al. 1985, 1990; Mammucari et al. 1988; Martin et al. 1990; Weddell 1989; Weddell et al. 1990; Caltagirone et al. 1989a; Blonder et al. 1993) also took into account the emotional valence of the stimulus to be appraised or of the expressed emotion in order to check the hypothesis of a different specialisation of the right and left hemisphere for negative and positive emotions respectively. However, in these studies no interaction between emotional valence and side of lesion was usually found.

Taken together, both results of experiments conducted in normal subjects and those carried out in patients with unilateral brain lesions are much more consistent with the predictions of 'right hemisphere dominance' than with those of the 'valence hypothesis'. However, some investigations conducted in normal subjects by Reuter-Lorenz and Davidson (1981), Reuter-Lorenz et al. (1983) and Natale et al. (1983) suggested an advantage of the right hemisphere in the recognition of negative emotions and of the left hemisphere in the recognition of positive emotions. Furthermore, Schwartz et al. (1979) recorded facial electromyography (EMG) in response to emotional questions and found that positive emotion questions elicited relatively greater right than left zygomatic muscle activity, whereas negative emotion questions elicited relatively greater left than right zygomatic muscle activity. They considered these results as consistent with the hypothesised specialisation of the left and right cerebral hemispheres for the mediation of positive and negative emotions, respectively.

Seemingly strong support for the 'valence hypothesis' was also provided by a paper in which Sackeim et al. (1982) gathered three retrospective studies aimed at exploring functional brain asymmetries in the regulation of emotion. In the first study, pathological laughing was associated with predominantly right-sided damage, whereas pathological crying was associated with predominantly left-sided lesions. In the second study, right hemispherectomy was associated with euphoric mood change, In the third study, a predominance of left-sided epileptic foci was found in patients with laughing (gelastic) epilepsy. These data, however, were not confirmed by further investigations (e.g. Feinstein et al. 1997; Wild et al. 2003; Arias 2011; Lauterbach et al. 2013) that aimed to clarify the pathophysiology and the lateralisation of pathological laughing and crying and of gelastic epilepsy.

4.3.3   Models of Emotional Laterality Prompted by Studies of the Nonverbal Communicative Aspects of Emotions

Three main models of emotional laterality have emerged from the above-mentioned studies of comprehension and expression of positive and negative emotions at the facial or vocal level. The first of these models, proposed by Ross and Mesulam (1979) and Ross (1981, 1984), assumed that the RH dominance for emotions may basically concern their nonverbal communicative aspects, rather than the autonomic components involved in emotional experience. The second model suggested that the greater hemiface asymmetry for negative, with respect to positive, emotions could be due to the interaction between the general RH superiority in the expression of spontaneous emotions and the leading role of the LH in the production of 'positive emotional expressions' (such as smiling) intentionally used for social communication. The third model, proposed by Davidson (1983, 1992, 1998) radically changed the perspective taken by previous interpretations. This model maintained that there are brain asymmetries (as measured by scalp recorded EEG activity) localised to the frontal region that are associated with the generation of emotion and that these hemispheric asymmetries are not related to the valence of the emotional stimulus but rather to the motivational (approach vs avoidance) system that is engaged by that stimulus. Each of these models will be taken separately into account in the present historical review.

4.3.3.1 The Hypothesis of a Right Hemisphere Dominance for Nonverbal Communication

Results of investigations which have consistently shown (both in healthy subjects and in brain-damaged patients) a general superiority of the right hemisphere for functions of emotional comprehension and expression have led some authors to hypothesise that the functional specialisation of the right hemisphere might primarily concern communicative aspects of emotions rather than other components. Following this line of thought, Ross (1981, 1984) and Blonder et al. (1991, 1993) suggested that the primary defect of right brain-damaged patients might be a disorder of nonverbal communication and that the emotional disturbances usually observed in these patients might simply be a byproduct of their inability to properly comprehend and express emotions. According to this viewpoint, the 'indifference reactions' of right brain-damaged patients should not be considered as an inappropriate form of emotional reaction but as a consequence of a basic inability to correctly evaluate emotional signals and to express an otherwise intact emotional experience. A basic objection can be raised with regard to this interpretation of the emotional disturbances of right brain-damaged patients based on results obtained studying the communicative aspects of emotions. In fact, if this hypothesis is correct, the emotional experience and the vegetative response associated with spontaneous emotional reactions should be intact in right brain-damaged patients, because their apparent indifference should essentially be due to an inability to express emotions. Results of investigations which have studied the autonomic components of emotions in normal subjects and in patients with unilateral brain damage (see Sect. 4.4 of this chapter) are, however, clearly inconsistent with this prediction.

4.3.3.2 The Hypothesis That Hemispheric Asymmetries May Concern Two Hierarchical Levels, Rather than Two Opposite Dimensions of Emotions

On the other hand, the second model, originally proposed by Buck (1984), Rinn (1984) and Etcoff (1986) and subsequently developed by Gainotti et al. (1993), Ross et al. (1994) and Mandal and Ambady (2004), suggests that hemispheric asymmetries might concern two hierarchical levels of emotions or of emotional processing rather than two different components of emotions. For example Etcoff (1986) attributed the less lateralised production of 'positive expressions' reported by some authors in normal subjects to the intentional use of smiling and other positive facial expression for communicative purposes. This author noted that smiling does not only differ from other facial emotional expressions because of its positive polarity, but also because it represents the easiest 'emotional' facial expression to reproduce voluntarily and the one most currently used for approach and social communication. In the case of smiling, LH dominance for the intentional control of the facial expressive apparatus could counterbalance the greater left half face expressiveness resulting from the overall RH prevalence for functions of emotional expression. Buck (1984) and Rinn (1984) also proposed that the greater asymmetry between the normal right and left hemiface in the expression of negative emotions could be due to the greater inhibition exerted by the left hemisphere on the right half face in the overt expression of these socially censurable emotions. Smiling could not present this asymmetry because it is not subject to this inhibition (and is on the contrary voluntarily used for social purposes) by the left hemisphere. The hypothesis of a left hemisphere dominance in the intentional control of the facial expressive apparatus is also supported by data obtained in three split-brain patients by Gazzaniga and Smylie (1990). These authors showed that although both hemispheres can generate spontaneous facial expressions only the left hemisphere can efficiently generate voluntary (smiling) expressions. Two different versions of this model, based respectively on the hierarchical distinctions between the 'primary/basic' and the 'social/complex' emotions and between the 'schematic' and 'conceptual' levels of emotional processing proposed by Leventhal (1979, 1987), were suggested respectively by Ross et al. (1994) and by Gainotti et al. (1993).

Ross et al. (1994) argued that the hypothesis of right hemisphere dominance for primitive emotions and left hemisphere prevalence for social emotions could account for results supporting both the 'right hemisphere hypothesis' and the 'valence hypothesis'. This statement is based on the reasoning that primitive emotions, which are lateralised to the right hemisphere, constitute the most emotional schemata and usually have a negative valence; by contrast, social emotions, which are lateralised to the left hemisphere, are less numerous and generally have a positive valence. Therefore, both studies that show an overall superiority of the right hemisphere for emotional functions and those that point to a different involvement of the right hemisphere in negative emotions and of the left hemisphere in positive emotions might at least in part be explained by this interpretation.

Gainotti et al. (1993) suggested that the right and left hemispheres might be preferentially involved in different hierarchical levels of emotions, i.e. that the right hemisphere might be more involved in the automatic and the left hemisphere in the controlled aspects of emotions. The RH might be mainly involved at the schematic level (where the 'emotional schemata', corresponding to spontaneous emotions, are automatically elicited and are accompanied by subjective feelings). By contrast, the LH might be mainly involved in the conceptual level, which contains a set of rules for responding appropriately (i.e. deliberately rather than spontaneously) to emotionally laden situations and plays a more important role in functions of intentional control of the emotional expressive apparatus.

4.3.3.3 The Hypothesis Which Assumes That Frontal Lobe Asymmetries Are More Related to the Motivational System Engaged by the Stimulus than to Its Emotional Valence

Results obtained in some investigations by studying the communicative aspects of emotions and originally considered as consistent with the 'valence hypothesis' were interpreted differently by Davidson (1983, 1992, 1998) based on electroencephalographic (EEG) asymmetries observed at the level of the frontal lobes in depressed individuals (Schaffer et al. 1983) and in normal subjects (Davidson et al. 1979; Davidson and Fox 1982), during the expression of positive and negative emotions. Davidson (1983, 1992, 1998) claimed that these asymmetries were not related to the valence of the emotional stimulus but rather to the motivational system that is engaged by that stimulus and proposed that the left prefrontal cortex (PFC) might be involved in a system inducing approach to appetitive stimuli and that the right PFC might be involved in a system that promotes withdrawal from aversive stimuli. Therefore, Davidson (1983) maintained that emotion is associated with left or right lateralisation according to the extent to which it is accompanied by approach or avoidance motivation. Furthermore, Davidson and Irwin (1999) also claimed that individual differences in approach- and withdrawal-related emotional reactivity and temperament are associated with stable differences in baseline measures of activation asymmetry in these anterior regions and that phasic state changes in emotion result in shifts in anterior activation asymmetries that are superimposed on these stable baseline differences. As the attention of neuroscientists at that time was focused on the physiology of emotion and the integration between emotion and cognition [and the term 'affective neuroscience', coined by Davidson and Sutton 1995, substantially contributed to the development of this research field], the measure of the frontal EEG asymmetries and the motivational approach-avoidance model became extremely popular, leading to many investigations (for reviews see Coan and Allen 2004; Harmon-Jones et al. 2010; Miller et al. 2013). However, some authors who recently provided a detailed review of research on the relationship between PFC and the approach/withdrawal vs the classical valence hypothesis acknowledge that this model is challenged by failures to find the expected asymmetries in response to affective images (e.g. Harmon-Jones et al. 2010). Other authors (e.g. Spielberg et al. 2008) noted that often this research does not distinguish emotional valence from motivational direction because most pleasant emotions are coupled with approach motivation and most unpleasant emotions are coupled with withdrawal motivation. Still other authors challenged the equivalence approach = positive and withdrawal = negative emotions, stressing the fact that anger, which is a negative emotion (Russell and Barrett 1999), often evokes an approach motivation and usually coincides with activation of the left hemisphere (Harmon-Jones and Allen 1998).

To disentangle the effects due to the valence of emotionally laden situations from those of their motivational components, some authors (e.g. Waldstein et al. 2000; Prete et al. 2015, 2018; Wyczesany et al. 2018) conducted experiments in which the valence hypothesis was checked by comparing stimuli that express respectively happiness and anger, because these emotions are representative of positive and negative valence, even if both are coupled with approach motivation. However, neither neuroimaging nor electrophysiological investigations were able to disentangle this issue by reporting alternative patterns of results, providing support for the 'right hemisphere', the 'valence' or the 'approach-avoidance' hypothesis.

More in general, it must be acknowledged that even though the motivational model has been very influential, for at least two reasons it cannot explain the results of experimental studies conducted in normal subjects and in patients with unilateral brain damage reported in Sect. 4.3 of the present chapter. The first reason is that the difference between positive and negative emotions studied at the expressive level in these investigations can only in part be captured by the distinction between approach and withdrawal tendencies proposed by Davidson (1983, 1992, 1998). The second reason is that the above-mentioned studies showed that the RH dominance not only concerns the expression of emotions, subtended by the anterior frontal regions, but also the recognition of emotionally laden information, based on the activity of the posterior cerebral cortices; by contrast, the approach/withdrawal model only concerns the experience and expression of emotions at the level of the PFC. In fact, a general dominance of the RH in the processing of emotional information was acknowledged by Davidson (1983), who suggested an interaction between the right/left and the anterior/posterior dichotomy in the regulation of human emotions. This interactive account assumed overall dominance of the right parietal region for the treatment of emotional information and a diverging specialisation of the right and left frontal lobes for the expression of negative and positive emotions. However, if this model was correct, we should predict that the interaction between valence and laterality will be greater at the expressive than at the receptive level; however, the data reported in Sect. 3.3 of this chapter have shown a trend in the opposite direction. A right hemisphere advantage for negative emotions and a left hemisphere advantage for positive emotions was, indeed, sometimes observed in receptive tasks (e.g. Dimond and Farrington 1977; Reuter-Lorenz and Davidson 1981; Reuter-Lorenz et al. 1983; Natale et al. 1983) but much less frequently in expressive tasks, where only Schwartz et al. (1979) found an opposite EMG asymmetric zygomatic muscle activity for negative vs positive emotions.

Furthermore, Miller et al. (2013) noted that the literature supporting valence or approach interpretations has often treated the left and right frontal lobes as single functional units. They argued that findings which appear contradictory at the level of the frontal lobes as units of analysis can be accommodated by the considerable differentiation, in specialisation and activation, of subregions of the frontal cortex, including their connectivity to each other and to other regions.

Finally, an even more radical objection to the motivational approach-avoidance model was expressed by authors who criticised both the relation between PFC and approach/avoidance tendencies (e.g. Spielberg et al. 2008) and the use of frontal EEG asymmetries to assess emotion or motivation. In particular, Allen et al. (2018) and Reznik and Allen (2018) recently showed that there is insufficient evidence in support of the relationships between frontal EEG asymmetries and approach- or withdrawal-related emotional reactions or the corresponding personality traits.

For these reasons, even though I acknowledge the importance of the motivational approach-avoidance model, there will be no more detailed discussions of results obtained following this line of research in subsequent sections of this volume.

4.4   Experimental and Clinical Investigations That Studied Laterality of the Autonomic Components of Emotions

As already noticed while discussing the early observations on the relationships between emotions and cerebral dominance, Hirschl and Potzl (quoted by Schilder 1935) hypothesised that greater involvement of the right hemisphere in vegetative functions could explain the Babinski's (1914) observation that anosognosia and anosodiaphoria are usually due to right hemisphere damage. This hypothesis, however, had not captured the attention of other scientists, and the assumption of a possible asymmetrical representation of autonomic functions was submitted to empirical study only after the period of the clinical studies which showed that emotions are lateralised in the human brain. The first study explicitly designed to investigate this issue in patients with unilateral brain damage was undertaken by Heilman et al. (1978), who studied the galvanic skin response to painful stimuli applied to the hand ipsilateral to the damaged hemisphere. These authors observed a flattened vegetative response in patients with right hemisphere lesions and, in particular, in those with emotional indifference and unilateral spatial neglect. After this seminal study, other investigations focused attention on hemispheric asymmetries for autonomic functions, following three different lines of research: (a) the psychophysiological correlates of emotional activation in patients with unilateral brain lesions; (b) hemispheric asymmetries in cardiac autonomic control; (c) the psychophysiological correlates of the selective emotional stimulation of the right or left hemisphere in normal subjects. Each of these lines of research will be taken into account separately in the present historical survey. A discussion of the possible different lateralisation of sympathetic vs parasympathetic activities and of the relationship between lateralisation of emotions and autonomic functions will conclude this section of the chapter dealing with the history of research on laterality of the autonomic components of emotions.

4.4.1   Psychophysiological Correlates of Emotional Activation in Unilateral Brain-Damaged Patients

Morrow et al. (1981), Zoccolotti et al. (1982) and Meadows and Kaplan (1994) studied the galvanic skin response to emotional and non-emotional slides in patients with unilateral brain lesions and in normal controls. Similar investigations were conducted by Zoccolotti et al. (1986) and Caltagirone et al. (1989b) who studied heart rate changes and galvanic skin response to emotional (positive and negative) and neutral short movies and by Andersson and Finset (1998) who studied stress reactivity (through heart rate, skin conductance level, and number of spontaneous skin conductance responses) in right and left brain-damaged patients. Furthermore, Làdavas et al. (1993) studied the physiological responses provoked by the subliminal projection of emotional slides to the right and left hemisphere in a split-brain patient. All of these studies showed an important reduction of the galvanic skin response and other indices of autonomic activation in patients with right brain lesions (and large physiological responses after subliminal projection of emotional slides to the intact right hemisphere); however, the vegetative response of left brain-damaged patients was much more variable, but usually lower than that of control subjects (see Hagemann et al. 2003 for review).

4.4.2   Hemispheric Asymmetries for Autonomic Heart Control

Since the seminal work of Mizeres (1958) and Levy et al. (1966), it has been known that at the level of the peripheral autonomic structures both sympathetic and parasympathetic innervation of the heart are strongly lateralised. Sympathetic outflow originates from the vasomotor centres of bulb and medulla oblongata, which project to the preganglionic neurons, situated in the intermediolateral cell columns of the spinal cord and innervate the heart through cells lying in the stellate ganglion (Levy and Martin 1990). Parasympathetic outflow originates from neurons of the nucleus ambiguous and of the dorsal motor nucleus of the vagus (both located in the medulla oblongata) and course down through the vagal nerve to intracardiac ganglia in the wall of the heart (Levy and Martin 1990). Both afferent (Katchanov et al. 1996) and efferent autonomic innervation of the heart (Wittling 1995) have been described as strongly asymmetric. In more recent years, however, investigations in human subjects have clearly shown that not only the peripheral innervation, but also the cerebral modulation of the heart's autonomic control is lateralised. These conclusions were reached studying: (1) the influence of unilateral brain lesions on cardiac functions, (2) the consequences of unilateral pharmacological hemispheric inactivation and (3) the cardiac effects of unilateral hemispheric stimulation.

1. Several authors studied the *cardiac effects of unilateral brain lesions* and showed that physiological heart rate variability (HRV) is reduced more by right than by left brain lesions. This observation was made in very different experimental conditions, spanning from the presentation of emotional stimuli (Zoccolotti et al. 1986; Caltagirone et al. 1989b) or of an attention demanding task (Yokoyama

et al. 1987; Andersson and Finset 1998) to the response to increased respiratory activity (Naver et al. 1996). Now, this reduction of HRV is certainly abnormal, since it is associated with an increased risk of sudden death in patients with and without a history of myocardial infarction (Kleiger and Miller 1987; Johnson and Robinson 1988). Tokgozoglu et al. (1999) and Colivicchi et al. (2004) showed that stroke in the region of the right insula leads to decreased HRV and to increased incidence of sudden death. Colivicchi et al. (2004) also documented a negative correlation between an important parameter of HRV (the SDNN) and all kinds of arrhythmia mortality in conditions of autonomic imbalance. Now, since a reduction of SDNN is deemed to reflect a diminished vagal activity directed to the heart and was mainly due to a right insular stroke in the Colivicchi et al.'s (2004) study, these data suggest a prevalent involvement of the right insula in parasympathetic functions. Consistent with these suggestions were results obtained by Naver et al. (1996) and Barron et al. (1994) in stroke patients and by Massetani et al. (1997) in patients with temporal lobe epilepsy (TLE). Naver et al. (1996) demonstrated that, compared with left-sided stroke and with control subjects, stroke location on the right side was associated with reduced respiratory heart rate variability (a reflex mainly under parasympathetic control). Barron et al. (1994), performing power spectrum analysis of relative risk (RR) variability in the electrocardiogram, showed that reduction of the cardiac para-sympathetic innervation was significantly greater after right-sided than left-sided stroke and using the same spectral analysis of RR variability to study alterations of cardiac functions in patients with TLE, Massetani et al. (1997) showed that disorders of autonomic cardiac control were more severe in cases with right EEG focus. On the other hand, Sander and Klingelhofer (1995) showed that in patients with hemispheric brain infarction changes of circadian blood pressure patterns and cardiovascular parameters indicate right lateralisation of sympathetic activation.

2. The *consequences of unilateral pharmacological brain inactivation on cardiac autonomic control* were studied by Rosen et al. (1982), Zamrini et al. (1990) and Yoon et al. (1997) in patients submitted to unilateral pharmacological inactivation with the Wada test. Injection of sodium amytal into the left carotid artery was followed by increased heart rate, considered as physiologically due to the stress provoked by the consequences of left hemisphere inactivation. However, Rosen et al. (1982) and Zamrini et al. (1990) did not observe a similar response after injection of sodium amytal into the right carotid artery. On the other hand, Yoon et al. (1997), tried to investigate the balance between sympathetic and parasympathetic inactivation, by studying the power spectral analysis of heart rate variability before and after intracarotid amytal injection. They observed a shift toward sympathetic predominance after left hemisphere inactivation, but no significant change of this balance after injection of sodium amytal into the right carotid artery. They concluded that the right hemisphere plays a leading role in sympathetic activities. This suggestion was confirmed by data obtained by Oppenheimer et al. (1996) in a series of patients with lesions confined to the left insula, because in these patients the cardiac autonomic balance was shifted toward a sympathetic predominance.

3. Also consistent with this suggestion were results obtained by Oppenheimer et al. (1992) studying heart rate and blood pressure changes during *electrical stimulation of the right and left insular cortex*. Sympathetic effects, with tachycardia and increased blood pressure were, in fact, mainly elicited by stimulation of the right insular cortex (see Oppenheimer 2006 for a general review of this issue).

4.4.3 Psychophysiological Correlates of the Selective Emotional Stimulation of the Right and Left Hemispheres in Normal Subjects

The psychophysiological correlates of the selective emotional stimulation of the right and left hemisphere were studied with different experimental procedures and with contrasting results in normal subjects by Wittling and coworkers (Wittling 1990, 1995; Wittling et al. 1998a, b) and by Spence et al. (1996). Wittling and coworkers used lateralised film presentation to investigate blood pressure changes (Wittling 1990), the power spectral analysis of heart rate variability (Wittling et al. 1998a) and various indices of myocardial activity (Wittling et al. 1998b). The high frequency band of the spectral power was taken as an index of parasympathetic activity, whereas measures of myocardial performance were used to evaluate the sympathetic influence on the heart. The authors observed a greater increase in blood pressure and measures of myocardial performance during right hemisphere film presentation, and a significant increase in the high frequency components of the spectral power during left hemisphere film presentation. They considered these findings as indicative of a different specialisation of the right hemisphere for sympathetic activities and of the left hemisphere for the parasympathetic control of the heart. Spence et al. (1996), on the other hand, showed normal subjects emotional and neutral slides that were briefly lateralised to the right or left visual field and used heart rate deceleration and pulse volume as measures of parasympathetic and sympathetic activation, respectively. The largest psychophysiological responses, concerning both pulse volume and heart rate deceleration, were obtained after presentation of the emotional slides to the right hemisphere. In contrast to Wittling's conclusions, Spence et al. (1996) claimed that both sympathetically mediated vasoconstriction and parasympathetically mediated heart rate deceleration are predominantly subserved by the right hemisphere. Therefore, the problem of the relationships between lateralisation of autonomic and emotional functions remains open and will be taken into account again in Chap. 6 and in the concluding remarks of the present volume.

4.5 Experimental and Clinical Investigations That Studied the Conscious Experience of Emotions

The relationships between autonomic response and subjective experience of emotions have been discussed since the James–Lange proposal (James 1884; Lange 1885) that physiological changes precede emotions, which are equivalent to our subjective experience of physiological changes, and are experienced as feelings, or conscious emotions (Zuckerman et al. 1981; Dalgleish 2004; Prinz 2004). Research carried out with normal subjects or brain-damaged patients suggests that the subjective experience provoked by the presentation of emotional stimuli might be linked more to the right than to the left hemisphere.

4.5.1 Research Conducted in Normal Subjects

Wittling and Roscmann (1993) showed films with emotional content to the right and left hemisphere of normal subjects using a technique for the lateralisation of visual input that allowed prolonged viewing while permitting free ocular scanning. Subjective emotional experience were assessed by means of a continuous rating of emotional arousal experienced during the movie as well as by retrospective ratings of ten different emotional qualities. Presenting both positive and negative films to the right hemisphere resulted in stronger subjective responses in the continuous emotion rating as well as in some retrospectively assessed ratings compared to left hemispheric presentation, which suggests higher responsiveness of the right hemisphere to subjective emotional experience.

4.5.2 Investigations Conducted in Patients with Unilateral Brain Lesions

Evidence gathered in unilateral brain-damaged patients in support of the hypothesis that the subjective experience of emotions may be linked more to the right than to the left hemisphere is less direct and mainly consists of an unexpected observation made by Mammucari et al. (1988) in their study on expressive facial reactions to the presentation of emotionally laden film clips (see Sect. 4.3 of this chapter). These authors noted that during the presentation of a very unpleasant film (showing a bloody surgical scene) normal subjects and patients with left hemisphere lesions often tended to look away from the screen, whereas right brain-damaged patients rarely presented gaze avoidance. Mammucari et al. (1988) hypothesised that control subjects and left brain-damaged patients tended to engage in visual avoidance behaviours because the crude scene made them anxious and emotionally disturbed. On the contrary, patients with right hemisphere lesions did not show a similar tendency to look away from the screen because they were less emotionally involved and thus more indifferent to the emotional situation. This hypothesis was checked by Caltagirone et al. (1989b), who studied the relations between presence of visual avoidance and changes of heart rate both in normal subjects and in patients with unilateral brain lesions. Results were consistent with the hypothesis, because in control subjects and in left brain-damaged patients the decrease in heart rate was significantly linked to the presence of avoidance eye movements, while in patients with right-sided lesions the changes were very slight and were not linked to the presence of gaze avoidance. We can conclude that results of investigations conducted both in healthy subjects and in patients with unilateral brain damage indicate that the right hemisphere is critically involved not only in tasks of nonverbal emotional communication but also in the generation of the autonomic components of the emotional response and of the concomitant subjective experience of emotions.

4.6   Possible Implications for Psychiatry of Different Models of Emotional Lateralisation

We have seen in Sect. 4.2 of this chapter that the different interpretations of the diverging emotional behaviour shown by right and left brain-damaged patients proposed by Terzian and Cecotto (1958), Alemà and Donini (1960) and Perria et al. (1961), and by Gainotti (1969, 1972) gave rise respectively to the 'valence hypothesis' and to the 'right hemisphere hypothesis'. These accounts, which are still considered the most important models in this area, allow making different predictions about the lateralisation of various aspects of psychopathology. In fact, the 'valence hypothesis' could suggest that inactivation of the left-sided network underpinning positive emotions causes a depressive syndrome with the characteristics of a major biological depression, whereas disruption of the right-sided structures subserving negative emotions could elicit a manic syndrome. These were, indeed, the main features of the 'depressive-catastrophic reaction' and of the 'euphoric-maniacal reaction' described by Terzian and Cecotto (1958), Alemà and Donini (1960) and Perria et al. (1961) after pharmacological inactivation of the left and of the right hemisphere. On the other hand, the 'right hemisphere hypothesis' might suggest that different kinds of psychopathology are subtended by a right hemisphere dysfunction, just as different kinds of abnormal emotional reactions shown by right brain-damaged patients were considered by Gainotti (1972) as pointing to a major involvement of the right hemisphere in emotional functions. Data supporting both the predictions based on the 'valence hypothesis' and those based on the 'right hemisphere hypothesis' can be found in the psychiatric literature. The first prediction could be supported by the observation that in stroke patients a major depression is usually observed in subjects with left frontal lesions (e.g. Robinson et al. 1984; Lipsey et al. 1986; Bhogal et al. 2004; Robinson 2006), whereas mania is mainly found in patients with right brain damage (e.g. Cummings and Mendez 1984; Robinson et al. 1988; Starkstein et al. 1990; Santos et al. 2011). On the other hand, the second prediction is supported by two sets of data: (a) studies questioning the relationships between major post-stroke depression and left frontal lesions (e.g. Gainotti et al. 1997, 1999; Carson et al. 2000) and between secondary mania and euphoric shift of the mood (e.g. Starkstein and Robinson 1997; Braun et al. 1999, 2008); (b) observations showing that the right hemisphere could be involved in a large array of psychopathological conditions, ranging from the lateral distribution of psychogenic pain and conversion reactions (e.g. Galin 1974; Stern 1977; Devinsky et al. 2001; Perez et al. 2012) to the anatomic correlates of delusional reduplication and misidentification (e.g. Feinberg and Shapiro 1989; Murai et al. 1997; Devinsky 2009; Feinberg 2013). However, a thorough discussion of these complex issues might be inappropriate here, because it would disrupt the line of reasoning that specifically concerns the links between emotions and brain laterality. This discussion will, therefore, be taken up in Chap. 6, which deals particularly with the psychopathological implications of hemispheric asymmetries for emotions.

References

Alemà G, Donini G. Sulle modificazioni cliniche ed elettroencefalografiche da introduzione intrac-arotidea di iso-amil-etil-barbiturato di sodio nell'uomo. Boll Soc Ital Biol Sper. 1960;36:900–4.

Allen JJB, Keune PM, Schönenberg M, Nusslock R. Frontal EEG alpha asymmetry and emotion: from neural underpinnings and methodological considerations to psychopathology and social cognition Psychophysiology. 2018;55 https://doi.org/10.1111/psyp.13028.

Andersson S, Finset A. Heart rate and skin conductance reactivity to brief psychological stress in brain-injured patients. J Psychosom Res. 1998;44:645–56.

Arias M. [Neurology of laughter and humour: pathological laughing and crying]. [article in Spanish]. Rev Neurol. 2011;53:415–21.

Asthana HS, Mandal MK. Hemifacial asymmetry in emotion expression. Behav Modif. 1998;22:177–83.

Asthana HS, Mandal MK. Visual-field bias in the judgment of facial expression of emotion. J Gen Psychol. 2001;128:21–9.

Babinski J. Contribution à l'étude des troubles mentaux dans l'hémiplégie organique cérébrale (Anosognosie). Rev Neurol. 1914;27:845–8.

Barron SA, Rogovski Z, Hemli J. Autonomic consequences of cerebral hemisphere infarction. Stroke. 1994;25:113–6.

Benowitz LI, Bear DM, Rosenthal R, Mesulam MM, Zaidel E, Sperry RW. Hemispheric special-ization in nonverbal communication. Cortex. 1983;19:5–11.

Bhogal SK, Teasell R, Foley N, Speechley M. Lesion location and poststroke depression: system-atic review of the methodological limitations in the literature. Stroke. 2004;35:794–802.

Blonder LX, Bowers D, Heilman KM. The role of the right hemisphere in emotional communica-tion. Brain. 1991;114:1115–27.

Blonder LX, Burns AF, Bowers D, Moore RW, Heilman KM. Right hemisphere facial expressivity during natural conversation. Brain Cogn. 1993;21:44–56.

Borod JC, Koff E, Lorch M, Nicholas M. Channels of emotional expression in patients with uni-lateral brain damage. Arch Neurol. 1985;42:345–8.

Borod JC, Koff E, Lorch M, Nicholas M. The expression and perception of facial emotion in brain-damaged patients. Neuropsychologia. 1986;24:169–80.

Borod JC, Kent J, Koff E, Martin C, Alpert M. Facial asymmetry while posing positive and negative emotions: support for the right hemisphere hypothesis. Neuropsychologia. 1988;26:759–64.

Borod JC, StClair J, Koff E, Alpert M. Perceiver and poser asymmetries in processing facial emo-tion. Brain Cogn. 1990;13:167–77.

Bowers D, Bauer RM, Coslett HB, Heilman KM. Processing of faces by patients with unilateral hemisphere lesions. I. Dissociation between judgments of facial affect and facial identity. Brain Cogn. 1985;4:258–72.

Brådvik B, Dravins C, Holtås S, Rosén I, Ryding E, Ingvar DH. Do single right hemisphere infarcts or transient ischaemic attacks result in aprosody? Acta Neurol Scand. 1990;81:61–70.

Braun CMJ, Larocque C, Daigneault S, Montour-Proulx I. Mania, pseudomania, depression, and pseudodepression resulting from focal unilateral cortical lesions. Neuropsychiatry Neuropsychol Behav Neurol. 1999;12:35–51.

Braun CMJ, Daigneault R, Gaudelet S, Guimond A. Diagnostic and statistical manual of mental disorders, fourth edition symptoms of mania: which one(s) result(s) more often from right than left hemisphere lesions? Compr Psychiatry. 2008;49:441–59.

Broca P. Sur la faculté du langage articulé. Bull Soc Antropol. 1865;6:377–93.

Bruyer R. Asymmetry of facial expression in brain damaged subjects. Neuropsychologia. 1981;19:615–24.

Bryden MP. Laterality: Hemispheric specialization in intact brain. New York: Academic Press; 1982.

Buchtel HA, Campari F, DeRisio C, Rota R. Hemispheric differences in discriminative reaction time to facial expression. Ital J Psychol. 1978;5:159–69.

Buck R. The communication of emotions. New York: Guilford Press; 1984.

Buck R, Duffy RJ. Non-verbal communication of affect in brain-damaged subjects. Cortex. 1980;16:351–62.

Caltagirone C, Ekman P, Friesen W, Gainotti G, Mammucari A, Pizzamiglio L, et al. Posed emotional expression in unilateral brain damaged patients. Cortex. 1989a;25:653–63.

Caltagirone C, Zoccolotti P, Originale G, Daniele A, Mammucari A. Autonomic reactivity and facial expression of emotions in brain-damaged patients. In: Gainotti G, Caltagirone C, editors. Emotions and the dual brain. Heidelberg: Springer; 1989b. p. 204–21.

Campbell R. Asymmetries in interpreting and expressing posed facial expression. Cortex. 1978;14:327–42.

Cancelliere AEB, Kertesz A. Lesion localization in acquired deficits of emotional expression and comprehension. Brain Cogn. 1990;13:133–47.

Carmon A, Nachshon I. Ear asymmetry in perception of emotional non-verbal stimuli. Acta Psychol. 1973;37:351–7.

Carson AJ, MacHale S, Allen K, Lawrie SM, Dennis M, House A, et al. Depression after stroke and lesion location: a systematic review. Lancet. 2000;356:122–6.

Cicone M, Wapner W, Gardner H. Sensitivity to emotional expressions and situations in organic patients. Cortex. 1980;16:145–58.

Coan JA, Allen JJB. Frontal EEG asymmetry as a moderator and mediator of emotion. Biol Psychol. 2004;67:7–50.

Colivicchi F, Bassi A, Santini M, Caltagirone C. Cardiac autonomic derangement and arrhythmias in right-sided stroke with insular involvement. Stroke. 2004;35:2094–8.

Critchley M. Personification of paralysed limbs in hemiplegics. Br Med J. 1955;30:284–6.

Critchley M. Observations on anosodiaphoria. Encéphale. 1957;46:540–6.

Cummings JL, Mendez MF. Secondary mania with focal cerebrovascular lesions. Am J Psychiatry. 1984;141:1084–7.

Dalgleish T. The emotional brain. Nat Rev Neurosci. 2004;5:583–9.

Darwin C. The expression of the emotions in man and animals. London: Murray; 1872. (reprinted Chicago: University of Chicago Press, 1965)

Davidson RJ. Hemispheric specialization for cognition and affect. In: Gale A, Edwards J, editors. Physiological correlates of human behavior. London: Academic Press; 1983. p. 203–26.

Davidson RJ. Anterior cerebral asymmetry and the nature of emotion. Brain Cogn. 1992;20:125–51.

Davidson RJ. Affective style and affective disorders: perspectives from affective neuroscience. Cogn Emot. 1998;12:307–30.

Davidson RJ, Fox NA. Asymmetrical brain activity discriminates between positive and negative stimuli in human infants. Science. 1982;218:1235–7.

Davidson RJ, Irwin W. The functional neuroanatomy of emotion and affective style. Trends Cogn Sci. 1999;3:11–21.

Davidson RJ, Sutton SK. Affective neuroscience: the emergence of a discipline. Curr Opin Neurobiol. 1995;5:217–24.

Davidson RJ, Schwartz GE, Saron C, Bennett J, Goleman DJ. Frontal versus parietal EEG asymmetry during positive and negative affect. Psychophysiology. 1979;16:202–3.

Dax M. Lésions de la moité gauche de l'encéphale coincidant avec l'oubli des signes de la pensée. Gaz Hebd Med Chir. 1865;2:259–60.

DeKosky ST, Heilman KM, Bowers D, Valenstein E. Recognition and discrimination of emotional faces and pictures. Brain Lang. 1980;9:206–14.

Devinsky O. Delusional misidentifications and duplications: right brain lesions, left brain delusions. Neurology. 2009;72:80–7.

Devinsky O, Mesad S, Alper K. Nondominant hemisphere lesions and conversion nonepileptic seizures. J Neuropsychiatry Clin Neurosci. 2001;3:367–73.

Dimond SJ, Farrington L. Emotional response to films shown to the right or left hemisphere of the brain measured by heart rate. Acta Psychol. 1977;41:255–60.

Dopson W, Beckwith B, Tucker DM, Bullard-Bates P. Asymmetry of facial expression in spontaneous emotion. Cortex 1984;20:243–52.

Duda PD, Brown J. Lateral asymmetry of positive and negative emotions. Cortex. 1984;20:253–61.

Ekman P. Universal and cultural differences in facial expression of emotion. In: Cole JR, editor. Nebraska symposium on motivation. Lincoln: University of Nebraska Press; 1972. p. 207–83.

Ekman P. Expression and the nature of emotion. In: Scherer K, Ekman P, editors. Approaches to emotion. Hillsdale, NJ: Erlbaum; 1984. p. 319–44.

Ekman P, Friesen WV. The facial action coding system. Palo Alto, CA: Consulting Psychologists Press; 1978.

Etcoff NL. Selective attention to facial identity and facial emotion. Neuropsychologia. 1984;22:281–95.

Etcoff NL. The neuropsychology of emotional expression. In: Goldstein G, Tarter RE, editors. Advances in clinical neuropsychology. New York: Plenum Press; 1986. p. 127–79.

Feinberg TE. Neuropathologies of the self and the right hemisphere: a window into productive personal pathologies Front Hum Neurosci 2013; doi: https://doi.org/10.3389/fnhum.2013.00472. eCollection 2013., 7

Feinberg TE, Shapiro RM. Misidentification–reduplication and the right hemisphere. Cognitive and Behav Neurol. 1989;2:39–48.

Feinstein A, Feinstein K, Gray T, O'Connor P. Prevalence and neurobehavioral correlates of pathological laughing and crying in multiple sclerosis. Arch Neurol. 1997;54:1116–21.

Gainotti G. Réaction catastrophiques et manifestations d'indifférence au cours des atteintes cérébrales. Neuropsychologia. 1969;7:195–204.

Gainotti G. Emotional behavior and hemispheric side of the lesion. Cortex. 1972;8:41–55.

Gainotti G. The meaning of emotional disturbances resulting from unilateral brain injury. In: Gainotti G, Caltagirone C, editors. Emotions and the dual brain. Heidelberg: Springer; 1989. p. 147–67.

Gainotti G. A historical review of investigations on laterality of emotions in the human brain. J Hist Neurosci. 2019;26:1–19.

Gainotti G, Caltagirone C, Zoccolotti P. Left/right and cortical subcortical dichotomies in the neuropsychological study of human emotions. Cogn Emot. 1993;7:71–93.

Gainotti G, Azzoni A, Gasparini F, Marra C, Razzano C. Relation of lesion location to verbal and nonverbal mood measures in stroke patients. Stroke. 1997;28:2145–9.

Gainotti G, Azzoni A, Marra C. Frequency, phenomenology and anatomical-clinical correlates of major post-stroke depression. Br J Psychiatry. 1999;175:163–7.

Galin D. Implications for psychiatry of left and right cerebral specialization. Arch Gen Psychiatry. 1974;31:572–83.

Gazzaniga MS, Smylie CS. Hemispheric mechanisms controlling voluntary and spontaneous facial expressions. J Cogn Neurosci. 1990;2:239–45.

Goldstein K. The organism: a holistic approach to biology, derived from pathological data in man. New York: American Books; 1939.

Gorelick PB, Ross ED. The aprosodias: further functional-anatomical evidence for the organisation of affective language in the right hemisphere. J Neurol Neurosurg Psychiatry. 1987;50:553–60.

Hagemann D, Waldstein SR, Thayer JF. Central and autonomic nervous system integration in emotion. Brain Cogn. 2003;52:79–87.

Hager JC, Ekman P. The asymmetry of facial actions is inconsistent with models of hemispheric specialization. Psychophysiology. 1985;22:307–18.

Haggard MP, Parkinson AM. Stimulus and task factors as determinants of ear advantages. Q J Exp Psychol. 1971;23:168–77.

Hansch EC, Pirozzolo FJ. Task relevant effects on the assessment of cerebral specialization for facial emotion. Brain Lang. 1980;10:51–9.

Harmon-Jones E, Allen JJB. Anger and frontal brain activity: EEG asymmetry consistent with approach motivation despite negative affective valence. J Pers Soc Psychol. 1998;74:1310–6.

Harmon-Jones E, Gable PA, Peterson CK. The role of asymmetric frontal cortical activity in emotion-related phenomena: a review and update. Biol Psychol. 2010;84:451–62.

Harrington A. Unfinished business: models of laterality in the nineteenth century. In: Davidson RJ, Hugdahl K, editors. Brain asymmetry. Cambridge, MA: MIT Press; 1995. p. 3–27.

Harris LJ. Early theory and research on hemispheric specialization. Schizophr Bull. 1999;25:11–40.

Hécaen H. Clinical symptomatology in right and left hemisphere lesion. In: Mountcastle VB, editor. Interhemispheric relations and cerebral dominance. Baltimore: Johns Hopkins; 1962. p. 215–43.

Heilman KM, Scholes R, Watson RT. Auditory affective agnosia. Disturbed comprehension of affective speech. J Neurol Neurosurg Psychiatry. 1975;38:69–72.

Heilman KM, Schwartz HD, Watson RT. Hypoarousal in patients with the neglect syndrome and emotional indifference. Neurology. 1978;28:229–32.

Heilman KM, Bowers D, Speedie L, Coslett HB. Comprehension of affective and nonaffective prosody. Neurology. 1984;34:917–21.

Heller W, Levy J. Perception and expression of emotion in right-handers and left-handers. Neuropsychologia. 1981;19:263–72.

Hugdahl K, Iversen PM, Johnsen BH. Laterality for facial expressions: does the sex of the subject interact with the sex of the stimulus face? Cortex. 1993;29:325–31.

Hughes C, Chan JL, Su MS. Aprosodia in Chinese patients with right cerebral hemisphere lesions. Arch Neurol. 1983;40:723–36.

James W. What is an emotion? Mind. 1884;9:188–205.

Johnson RH, Robinson BJ. Mortality in alcoholics with autonomic neuropathy. J Neurol Neurosurg Psychiatry. 1988;51:476–80.

Katchanov G, Xu J, Hurt CM, Pelleg A. Electrophysiological-anatomic correlates of ATP-triggered vagal reflex in the dog. III. Role of cardiac afferents. Am J Phys. 1996;270:H1785–90.

Kinsbourne M, Bemporad B. Lateralization of emotion: a model and the evidence. In: Fox N, Davidson RJ, editors. The psychobiology of affective development. Hillsdale, NJ: Erlbaum; 1984. p. 353–81.

Kleiger RE, Miller P. Bigger JT, Moss AJ. Multicenter post infarction group: decreased heart rate variability and its association with increased mortality after acute myocardial infarction. Am J Cardiol. 1987;59:256–62.

Kolb B, Milner B. Observations on spontaneous facial expression after focal cerebral excisions and after intracarotid injection of sodium amytal. Neuropsychologia. 1981;19:505–14.

Kolb B, Taylor L. Affective behavior in patients with localized cortical excisions: role of lesion site and side. Science. 1981;214(4516):89–91.

Làdavas E, Umiltà C, Ricci-Bitti PE. Evidence for sex differences in right-hemisphere dominance for emotions. Neuropsychologia. 1980;18:361–6.

Làdavas E, Cimatti D, Del Pesce M, Tozzi G. Emotional evaluation with and without conscious stimulus identifications: evidence from a split-brain patient. Cogn Emot. 1993;7:95–114.

Landis T, Assal G, Perret E. Opposite cerebral hemispheric superiorities for visual associative processing of emotional facial expressions and objects. Nature. 1979;278:739–40.

Lange C. The emotions: a psychophysiological study. In: Dunlap E, editor. The emotions. Baltimore: Lippincott, Williams & Wilkins; 1885. p. 33–90.

Lauterbach EC, Cummings JL, Kuppuswamy PS. Toward a more precise, clinically--informed pathophysiology of pathological laughing and crying. Neurosci Biobehav Rev. 2013;37:1893–916.

Leventhal H. A perceptual- motor processing model of emotion. In: Pliner P, Blankestein K, Spiegel IM, editors. Perception of emotion in self and others, vol. 5. New York: Plenum; 1979. p. 1–46.

Leventhal H. A perceptual motor theory of emotion. In: Berkowitz L, editor. Advances in experimental social psychology, vol. 17. New York: Academic Press; 1987. p. 117–82.

Levy MN, Martin PJ. Neural control of the heart. In: Berne RM, Sperelakis N, editors. Handbook of physiology: circulation. Sect.2, The cardiovascular system, vol. 1. Bethesda, MD: American Physiological Society; 1990. p. 581–620.

Levy MN, Ng ML, Zieske H. Functional distribution of the peripheral cardiac sympathetic pathways. Circ Res. 1966;19:650–61.

Ley RG, Bryden HP. Hemispheric differences in processing emotions and faces. Brain Lang. 1979;7:127–38.

Ley RG, Bryden HP. A dissociation of right and left hemispheric effects for recognizing emotional tone and verbal content. Brain Cogn. 1982;1:3–9.

Lipsey JR, Spencer WC, Rabins PV, Robinson RG. Phenomenological comparison of poststroke depression and functional depression. Am J Psychiatry. 1986;143:527–9.

Luys JB. Etudes sur le dédoublement des opérations cérébrales et sur le rôle isolé de chaque hémisphère dans les phénemènes de la pathologie mentale. Bull Acad Natl Méd. 1879;8:516–34. 547-65

Luys JB. Traité clinique et pratique des maladies metale. Paris: Delahaye et Lecrosnier; 1881.

Luys JB. Le traitement de la folie. Paris: Rueff; 1893.

Mammucari A, Caltagirone C, Ekman P, Friesen W, Gainotti G, et al. Spontaneous facial expression of emotions in brain-damaged patients. Cortex. 1988;24:521–33.

Mandal MK, Ambady N. Laterality of facial expressions of emotion: universal and culture-specific influences. Behav Neurol. 2004;15:23–34.

Mandal MK, Singh SK. Lateral asymmetry in identification and expression of facial emotions. Cogn Emot. 1990;4:61–70.

Mandal MK, Asthana HS, Tandon SC. Judgement of facial expression of emotion in unilateral brain-damaged patients. Arch Clin Neuropsychol. 1993;8:171–83.

Martin C, Borod JC, Alpert M, Brozgold A, Welkowitz J. Spontaneous expression of facial emotion in schizophrenic and right brain-damaged patients. J Comm Dis. 1990;23:287–301.

Massetani R, Strata G, Galli R, Gori S, Gneri C, Limbruno U, et al. Alteration of cardiac function in patients with temporal lobe epilepsy: different roles of EEG-ECG monitoring and spectral analysis of RR variability. Epilepsia. 1997;38:363–9.

Meadows ME, Kaplan RF. Dissociation of autonomic and subjective responses to emotional slides in right hemisphere damaged patients. Neuropsychologia. 1994;32:847–56.

Miller GA, Crocker LD, Spielberg JM, Infantolino ZP, Heller W. Issues in localization of brain function: the case of lateralized frontal cortex in cognition, emotion, and psychopathology. Front Integr Neurosci. 2013 Jan 30;7:2. https://doi.org/10.3389/fnint.2013.00002.

Mills CK. The cerebral mechanisms of emotional expression. Trans Stud Coll Physicians Phila. 1912;34:381–90.

Mizeres NJ. The origin and course of cardioaccelerator fibers in the dog. Anat Rec. 1958;132:261–79.

Moreno CR, Borod JC, Welkowitz J, Alpert M. Lateralization for the expression and perception of facial emotion as a function of age. Neuropsychologia. 1990;28:199–209.

Morrow L, Vrtunski B, Kim Y, Boller F. Arousal response to emotional stimuli and laterality of lesion. Neuropsychologia. 1981;19:65–71.

Moscovitch M, Olds J. Asymmetries in spontaneous facial expressions and their possible relation to hemispheric specialization. Neuropsychologia. 1982;20:71–81.

Murai T, Toichi M, Sengoku A, Miyoshi K, Morimune S. Reduplicative paramnesia in patients with focal brain damage. Neuropsychiatry Neuropsychol Behav Neurol. 1997;10:190–6.

Natale M, Gur RE, Gur RC. Hemispheric asymmetries in processing emotional expressions. Neuropsychologia. 1983;21:555–65.

Naver HK, Blomstrand C, Wallin G. Reduced heart rate variability after right-sided stroke. Stroke. 1996;27:247–51.

Nicholls ME, Wolfgang BJ, Clode D, Lindell AK. The effect of left and right poses on the expression of facial emotion. Neuropsychologia. 2002;40:1662–5.

Oppenheimer SM. Cerebrogenic cardiac arrhythmias: cortical lateralization and clinical significance. Clin Auton Res. 2006;16:6–11.

Oppenheimer SM, Gelb A, Girvin JP, Hachinski VC. Cardiovascular effects of human insular cortex stimulation. Neurology. 1992;42:1727–32.

Oppenheimer SM, Kedem G, Martin WM. Left-insular cortex lesions perturb cardiac autonomic tone in humans. Clin Auton Res. 1996;6:131–40.

Peper M, Irle E. The decoding of emotional concepts in patients with focal cerebral lesions. Brain Cogn. 1997;34:360–87.

Perez DL, Barsky AJ, Daffner K, Silbersweig DA. Motor and somatosensory conversion disorder: a functional unawareness syndrome? J Neuropsychiatry Clin Neurosci. 2012;24:141–51.

Perria L, Rosadini G, Rossi GF. Determination of side of cerebral dominance with Amobarbital. Arch Neurol. 1961;4:173–81.

Prete G, Capotosto P, Zappasodi F, Laeng B, Tommasi L. The cerebral correlates of subliminal emotions: an eleoencephalographic study with emotional hybrid faces. Eur J Neurosci. 2015;42:2952–62.

Prete G, Capotosto P, Zappasodi F, Tommasi L. Contrasting hemispheric asymmetries for emotional processing from event-related potentials and behavioral responses. Neuropsychology. 2018;32:317–28.

Prigatano GP, Pribram KH. Perception and memory of facial affect following brain injury. Percept Mot Skills. 1982;54:859–69.

Prinz J. Emotions embodied. In: Solomon R, editor. Thinking about feeling: contemporary philosophers on emotions. Oxford: Oxford University Press; 2004. p. 44–59.

Reuter-Lorenz P, Davidson RJ. Differential contribution of the two cerebral hemispheres to the perception of happy and sad faces. Neuropsychologia. 1981;19:609–13.

Reuter-Lorenz P, Givis RP, Moscovitch M. Hemispheric specialization and the perception of emotion: evidence from right –handers and from inverted and noninverted left handers. Neuropsychologia. 1983;21:687–92.

Reznik SJ, Allen JJB. Frontal asymmetry as a mediator and moderator of emotion: an updated review. Psychophysiology. 2018; https://doi.org/10.1111/psyp.12965.

Rinn WE. The neuropsychology of facial expression: a review of the neurological and psychological mechanisms for producing facial expressions. Psychol Bull. 1984;95:52–77.

Robinson RG. The clinical neuropsychiatry of stroke. 2nd ed. Cambridge: Cambridge University Press; 2006.

Robinson RG, Kubos KL, Starr LB, Rao K, Price TR. Mood disorders in stroke patients. Importance of location of lesion. Brain. 1984;107:81–93.

Robinson RG, Boston JD, Starkstein SE, Price TR. Comparison of mania and depression after brain injury: causal factors. Am J Psychiatry. 1988;145:172–8.

Rosen AD, Gur RC, Sussman N, Gur RE, Hurtig H. Hemispheric asymmetry in the control of heart rate. Abstr Social Neurosci. 1982;8:917.

Ross ED. The aprosodias. Functional-anatomic organization of the affective components of language in the right hemisphere. Arch Neurol. 1981;38:561–9.

Ross ED. Right's hemisphere role in language. affective behaviour and emotion Trends Neurosci. 1984;7:342–6.

Ross ED, Mesulam MM. Dominant language functions of the right hemisphere? Prosody and emotional gesturing Arch Neurol. 1979;36:144–8.

Ross ED, Monnot M. Neurology of affective prosody and its functional-anatomic organization in right hemisphere. Brain Lang. 2008;104:51–74.

Ross ED, Homan RW, Buck R. Differential hemispheric lateralization of primary and social emotions: implication for developing a comprehensive neurology for emotions, repression. and the subconscious Neuropsychiatry Neuropsychol Behav Neurol. 1994;7:1–19.

Ross ED, Thompson RD, Yenkosky JP. Lateralization of affective prosody in brain and the callosal integration of hemispheric language functions. Brain Lang. 1997;56:27–54.

Rossi GF, Rosadini G. Experimental analysis of cerebral dominance in man. In: Millikan CJ, Darly FL, editors. Brain mechanisms underlying speech and language. New York: Grune and Stratton; 1967.

Russell JA, Barrett LF. Core affect, prototypical emotional episodes, and other things called emotion: dissecting the elephant. J Pers Soc Psychol. 1999;76:805–19.

Sackeim HA, Grega DM. Perceiver bias in the processing of deliberately asymmetric emotional expressions. Brain Cogn. 1987;6:464–73.

Sackeim HA, Gur RC. Lateral asymmetry in intensity of emotional expression. Neuropsychologia. 1978;16:473–81.

Sackeim HA, Gur RC, Saucy MC. Emotions are expressed more intensely on the left side of the face. Science. 1978;202:434–6.

Sackeim HA, Greenberg MS, Weiman AL, Gur RC, Hungerbuhler JP, Geschwind N. Hemispheric asymmetry in the expression of positive and negative emotions. Neurologic evidence Arch Neurol. 1982;39:210–8.

Safer MA, Leventhal H. Ear differences in evaluating emotional tones of voice and verbal content. J Exp Psychol Hum Percept Perform. 1977;3:75–82.

Sander D, Klingelhofer J. Changes of circadian blood pressure patterns and cardiovascular parameters indicate lateralization of sympathetic activation following hemispheric brain infarction. J Neurol. 1995;242:313–8.

Santos CO, Caeiro L, Ferro JM, Figueira ML. Mania and stroke: a systematic review. Cerebrovasc Dis. 2011;32:11–21.

Schaffer CE, Davidson RJ, Saron C. Frontal and parietal electroencephalogram asymmetry in depressed and nondepressed subjects. Biol Psychiatry. 1983;18:753–62.

Schilder P. The image and appearance of the human body. London: K. Paul; 1935.

Schlanger BB, Schlanger P, Gerstman L. The perception of emotionally toned sentences by right-hemisphere damaged and aphasic patients. Brain Lang. 1976;3:396–403.

Schmitt JJ, Hartje W, Willmes K. Hemispheric Asymmetry in the recognition of emotional attitude conveyed by facial expression, prosody and propositional speech. Cortex. 1997;33:65–81.

Schwartz GE, Ahern GL, Brown SL. Lateralized facial muscle response to positive and negative emotional stimuli. Psychophysiology. 1979;16:561–71.

Spence S, Shapiro D, Zaidel E. The role of the right hemisphere in the physiological and cognitive components of emotional processing. Psychophysiology 1996;33:112–22.

Spielberg JM, Stewart JL, Levin RL, Miller GA, Heller W. Prefrontal cortex, emotion, and approach/withdrawal motivation. Soc Pers Psychol Compass. 2008;2:135–53.

Starkstein SE, Robinson RG. Mechanism of disinhibition after brain lesions. J Nerv Ment Dis. 1997;185:108–14.

Starkstein SE, Mayberg HS, Berthier ML, Fedoroff P, Price TR, Dannals RF, et al. Mania after brain injury: neuroradiological and metabolic findings. Ann Neurol. 1990;27:652–9.

Stern DB. Handedness and the lateral distribution of conversion reactions. J Nerv Ment Dis. 1977;164:122–8.

Strauss E, Moscovitch M. Perception of facial expression. Brain Lang. 1981;13:308–32.

Suberi M, McKeever WF. Differential right hemispheric memory storage of emotional and non-emotional faces. Neuropsychologia. 1977;15:757–68.

Terzian H. Behavioural and EEG effects of intracarotid sodium amytal injection. Acta Neurochir. 1964;12:230–9.

Terzian H, Cecotto S. Su un nuovo metodo per la determinazione e lo studio della dominanza emisferica. Giornale di Psichiatria e Neuropatologia. 1958;87:889–924.

Tokgozoglu SL, Batur MK, Topçuoglu MA, Saribas O, Kes S, Oto A. Effects of stroke localization on cardiac autonomic balance and sudden death. Stroke. 1999;30:1307–11.

Tomkins SS. Affect, imagery, consciousness: vol. 1, the positive affects. New York: Springer; 1962.

Tomkins SS. Affect, imagery, consciousness: vol. 2, the negative affects. New York: Springer; 1963.

Tucker DM, Watson RT, Heilman KM. Discrimination and evocation of affectively intoned speech in patients with right parietal disease. Neurology. 1977;27:947–50.

Waldstein SR, Kop WJ, Schmidt LA, Haufler AJ, Krantz DS, Fox NA. Frontal electrocortical and cardiovascular reactivity during happiness and anger. Biol Psychol. 2000;55:3–23.

Weddell RA. Recognition memory for emotional facial expressions in patients with focal cerebral lesions. Brain Cogn. 1989;11:1–17.
Weddell RA, Miller JD, Trevarthen C. Voluntary emotional facial expressions in patients with focal cerebral lesions. Neuropsychologia. 1990;28:49–60.
Wild B, Rodden FA, Grodd W, Ruch W. Neural correlates of laughter and humour. Brain. 2003;126:2121–38.
Wittling W. Psychophysiological correlates of human brain asymmetry: blood pressure changes during lateralized presentation of an emotionally laden film. Neuropsychologia. 1990;28:457–70.
Wittling W. Brain asymmetry in the control of autonomic-physiologic activity. In: Asymmetry B, editor. Davidson RJ, Hugdahl K. Cambridge: MIT Press; 1995. p. 305–57.
Wittling W, Roscmann R. Emotion-related hemisphere asymmetry: subjective emotional responses to laterally presented films. Cortex. 1993;29:431–48.
Wittling W, Block A, Genzel S, Schweiger E. Hemisphere asymmetry in parasympathetic control of the heart. Neuropsychologia. 1998a;36:461–8.
Wittling W, Block A, Schweiger E, Genzel S. Hemisphere asymmetry in sympathetic control of the human myocardium. Brain Cogn. 1998b;38:17–35.
Wyczesany M, Capotosto P, Zappasodi F, Prete G. Hemispheric asymmetries and emotions: evidence from effective connectivity. Neuropsychologia. 2018;121:98–105.
Wylie DR, Goodale MA. Left-sided oral asymmetries in spontaneous but not posed smiles. Neuropsychologia. 1988;26:823–32.
Yokoyama K, Jennings R, Ackles P, Hood BS, Boller F. Lack of heart rate changes during attention-demanding tasks after right hemisphere lesions. Neurology. 1987;37:624–30.
Yoon BW, Morillo CA, Cechetto DF, Hachinski V. Cerebral hemispheric lateralization in cardial autonomic control. Arch Neurol. 1997;54:741–4.
Zamrini EY, Meador KJ, Loring DW, Nichols FT, Lee GP, Figueroa RE, et al. Unilateral cerebral inactivation produces differential left/right heart rate responses. Neurology. 1990;40:1408–11.
Zenhausern R, Notaro J, Grosso J, Schiano P. The interaction of hemispheric preference, laterality, and sex in the perception of emotional tone and verbal content. Int J Neurosci. 1981;13:121–6.
Zoccolotti P, Scabini D, Violani V. Electrodermal responses in patients with unilateral brain damage. J Clin Neuropsychol. 1982;4:143–50.
Zoccolotti P, Caltagirone C, Benedetti N, Gainotti G. Perturbation des réponses végétatives aux stimuli émotionnels au cours des lésions hémisphériques unilatérales. L'Encéphale. 1986;12:263–8.
Zuckerman M, Klorman R, Larrance DT, Spiegel NH. Facial, autonomic, and subjective components of emotion: the facial feedback hypothesis versus the externalizer–internalizer distinction. J Pers Soc Psychol. 1981;41:929–44.

Recent Trends in the Study of the Links Between Emotions and Brain Laterality

5

Contents

Results obtained recently by research directions that tried to clarify the relationships between emotions and hemispheric asymmetries have not been very consistent. One of these lines of research, i.e. the whole brain neuroimaging study of structures activated by emotional tasks in normal subjects, provided rather negative results. On the contrary, very interesting results were obtained by anatomo-clinical and activation studies that investigated laterality effects in structures that have a critical role in specific components of emotions and by anatomo-clinical and experimental

© Springer Nature Switzerland AG 2020
G. Gainotti, *Emotions and the Right Side of the Brain*,
https://doi.org/10.1007/978-3-030-34090-2_5

investigations that studied the emotional and behavioural disorders of patients with asymmetrical forms of fronto-temporal degeneration. Each of these lines of research will, therefore, be taken into account separately in this chapter.

5.1   Negative Data Obtained by Whole Brain Functional Neuroimaging Studies of the Neural Correlates of Emotions

In recent years the links between emotions and brain laterality have been intensively investigated using positron emission tomography (PET) and functional magnetic resonance imaging (fMRI) to study the activation produced in various brain structures of healthy subjects by experimental paradigms assessing different facets of emotions. To make a general survey of these whole brain activation studies, Wager et al. (2003) performed a quantitative meta-analysis of 65 neuroimaging studies of emotion. They focused on hypotheses concerning the right hemisphere dominance in emotional functions and the effects of emotional valence on regional brain activations. These authors failed to support both the 'right hemisphere' and the 'valence' hypothesis and similar results were obtained by Fusar-Poli et al. (2009) in a voxel-based meta-analysis of studies employing emotional faces paradigms in healthy subjects. Even more recently, Lindquist et al. (2016) tried to evaluate the brain basis of positive and negative affect with a meta-analysis of 397 fMRI and PET investigations. Their results failed to confirm the hypothesis assuming that independent brain regions may support positive and negative emotions. Levenson et al. (2014) noted, however, that these disappointing results should be considered with caution, because they could be due to several methodological and experimental factors. For instance, constraints and limits imposed by functional imaging methodologies prevent a full-blown study of emotional processes, because even if it were possible to induce powerful emotions in the scanner, the attendant muscular activity in the body and the face would produce huge signal artefacts. For this reason, a large amount of research conducted in the scanner is more concerned with the domain of emotional cognition (where the focus is on how we think about emotions and make judgments about them) than the actual processes as they unfold in real time. As for the confounding factors linked to the experimental condition, we can list those due to structural brain asymmetries (Watkins et al. 2001), to the different levels of emotional processing involved by the task, to the verbal or nonverbal nature of the stimuli used in different experiments and to control conditions, imaging data analysis and statistical thresholds. Furthermore, even the consistency of results obtained in more homogeneous studies is challenged by data obtained by Eugène et al. (2003), studying the impact of individual differences on the neural circuitry underlying sadness. These authors conducted two methodologically identical fMRI studies to identify the neural correlates of unhappiness. In the first of these studies, sadness was correlated with significant loci of activation in the anterior temporal pole and insula; in the second study, sadness was correlated with significant activation in the orbitofrontal and medial prefrontal cortices. Moreover, individual statistical parametric maps revealed a marked degree of inter-individual variability in both studies. It is also worth noting that generally imaging researchers have not looked for asymmetries and have rarely analysed data in a way that directly compares left and right activations (In contrast, for example to language researchers who have long used appropriate statistical methods to assess asymmetry).

Therefore, two sets of data can be considered more appropriate than results obtained in heterogeneous whole brain activation studies. The first refers to results obtained in anatomo-clinical or activation studies of brain structures such as the amygdala, the ventromedial prefrontal cortex (vmPFC) and the anterior insula, which are known to have very relevant and specific roles in emotional functions. The functional interactions between these components of the emotional system are schematically reported in Fig. 5.1.

The second set of data refers to results obtained studying emotional and behavioural disorders of patients with asymmetrical forms of frontal or temporal variants of fronto-temporal degeneration.

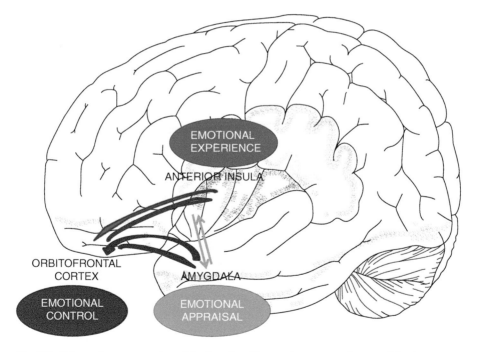

Fig. 5.1 Schematic representation of the functional interactions among the amygdala, the ventromedial prefrontal cortex and the anterior insula. Bilateral connections exist between the amygdala (where emotional information is computed), the orbitofrontal cortex (where they are integrated with social rules that allow emotional expression) and the anterior insular cortex (which contributes to the conscious experience of emotions). "Reproduced from Emotions and the Right Hemisphere: Can New Data Clarify Old Models? Guido Gainotti, The Neuroscientist, 25(3), 2019, pp. 258–270, Copyright © 2019, © SAGE Publications, with Permission"

5.2   Laterality Effects in Brain Structures That Have a Different Role in Emotional Functions

In recent years, several anatomo-clinical investigations with different experimental paradigms and studying different kinds of emotions tried to determine whether laterality effects could be found in brain structures that have a different role in emotional functions. For instance, Meletti et al. (2003, 2009) studied facial emotion recognition in subjects with symptomatic epilepsy, to evaluate whether anteromedial temporal lobe regions, particularly the right and left amygdala, participate in the recognition of emotions from facial expressions. They showed that: (a) subjects with right-sided mesial temporal sclerosis (MTS), but not those with left-sided MTS or other temporal or extratemporal epileptogenic lesions, were impaired in emotion recognition; (b) these patients were maximally impaired in the recognition of fear and to a lesser extent in the recognition of sadness and disgust. Even more recently, Tippett et al. (2018) adopted lesion-mapping techniques in individuals with acute right hemisphere stroke to investigate lesions associated with impaired recognition of prototypic emotional faces. In agreement with the 'right hemisphere' model of emotional laterality, they showed that right hemisphere stroke patients were significantly less accurate than controls on a test of facial recognition of both positive and negative emotions. However, they also showed that not all patients with right hemisphere lesions were impaired in emotional recognition, because only those with focal lesions in the amygdala or the anterior insula, were significantly impaired in the recognition of emotions from faces. Furthermore, the percentage of damage to the right temporal pole and orbitofrontal cortex contributed independently to predicting lower accuracy in recognising happy, surprised and neutral faces, consistent with the role of these areas in emotional recognition from faces reported in other studies with different aetiology of lesions. Consistent with the hypothesis of a critical role of right hemisphere structures in emotional processing are also data reported in Chap. 3 (Sect. 3.3) of this volume, which have shown that lesions sparing visual cortex and amygdala, but involving somatosensory and insular cortices can contribute to impair emotional face recognition (Adolphs et al. 2000). To explain these findings, Adolphs et al. (2000) assumed that viewing facial emotional expressions may trigger the corresponding emotional response in the perceiver and that somatosensory and insular cortices may detect these somatic and vegetative changes, providing information about the person's emotional state. Data reported by Adolphs et al. (2000) show, however, that this phenomenon only concerns the right somatosensory and insular cortices, suggesting a critical role of the right hemisphere even in these more complex forms of emotional processing. In any case, the meaning of laterality effects in brain structures that have a critical role in specific emotional functions was clarified more by results obtained on activation studies than by the above mentioned anatomo-clinical data (see Gainotti 2019a for survey).

5.2.1   Different Role of Right and Left Amygdala in the Evaluation of Emotional Stimuli

Morris et al. (1998) used PET to study the mechanism of an unconscious form of emotional learning in which an aversively conditioned masked emotional face elicited an unconscious emotional response. They showed that the masked presentation of the emotional face provoked a significant neural response in the right but not the left amygdala, whereas the unmasked presentation of the same stimulus enhanced neural activity in the left more than in the right amygdala. The authors concluded that the right amygdala has a major role in unconscious and the left in conscious forms of emotional learning. In a further paper, the same authors (Morris et al. 1999) tried to clarify the mechanisms through which this unconscious emotional learning can be mediated. To do this, on one hand they took into account the crucial role of the amygdala in classical emotional conditioning (LeDoux 1996), and on the other hand the existence of a cortical and a subcortical route (Papez 1937; LeDoux et al. 1986) through which perceptual stimuli might reach the amygdala (see Sect. 3.1 of Chap. 3). An increased correlation was observed between the right amygdala, pulvinar and superior colliculus (SC) during the unconscious (masked) presentation of conditioned emotional stimuli; however, no masking-dependent changes in correlation were observed among the same subcortical structures and the left amygdala. Morris et al. (1999) concluded that emotionally laden stimuli can be detected, processed and learned without conscious awareness by a right hemisphere subcortical pathway, mediating unconscious emotional learning. In line with this model, Wright et al. (2003) and Gläscher and Adolphs (2003) proposed that the right amygdala might be involved in the rapid detection of emotional stimuli, and that the left amygdala could play a role in the more elaborate stimulus evaluation. Results obtained by Morris et al. (1998, 1999) were confirmed with different neuroimaging techniques by Nomura et al. (2004), Noesselt et al. (2005) and Williams et al. (2006) and with electroencephalographic (EEG) or magnetoencephalographic (MEG) investigations by Balconi and Lucchiari (2008), Pegna et al. (2008), Luo et al. (2009) and Hung et al. (2010). All of these studies confirmed that the amygdala processes threat-related information through a right fast subcortical route and a left slower cortical feedback mechanism (see Gainotti 2012 for a more detailed description of these investigations). Results consistent with the hypothesis of right amygdala unconscious evaluation of emotional stimuli were also obtained by other authors who used different clinical or experimental paradigms. For example Williams and Mattingley (2004) showed that the emotion conveyed by a non-threatening emotional face in the extinguished hemifield can influence judgements of the emotion of a subsequent target face presented foveally. Analogous results were obtained by Schepman et al. (2018) in a dichotic listening study in which participants were asked to rate the stimuli's pleasantness in the attended to-be-rated or the unattended channel. When participants rated neutral stimuli and the unattended channel was positive/negative, the valence of the unattended channel affected the neutral ratings more strongly with left ear (right hemisphere, RH) processing of the affective sound. Both of these studies confirm (indicate) that the RH might be specialised in the unconscious processing of emotion via subcortical routes. In recent years, this hypothesis was also supported by Burra et al. (2019),

Cecere et al. (2014) and Bertini et al. (2019), who studied aspects of the affective blindsight and by Koller et al. (2019), who used probabilistic diffusion tractography to provide direct evidence for a subcortical pathway that transmits visual stimuli signalling threat, via the SC and pulvinar, from the retina to the amygdala. Burra et al. (2019) studied a patient with bilateral cortical blindness and affective blindsight using an fMRI paradigm in which fearful and neutral expressions were presented; they used faces that were either unfiltered or filtered to remove high or low spatial frequencies. Unfiltered fearful faces produced right amygdala activation, although the patient was unaware of the presence of the stimuli, suggesting that affective blindsight could rely on the right subcortical colliculo-pulvino-amygdalar pathway Cecere et al. (2014) and Bertini et al. (2019) showed that in hemianopic patients unseen fearful faces are implicitly processed and can facilitate the visual analysis of facial stimuli presented in the intact field. However, this facilitation is observed only in patients with left-sided lesions (and an intact right visual pathway), supporting the hypothesis that only the right hemisphere mediates implicit visual processing of fear signals. This hypothesis was also confirmed by Koller et al. (2019), who found that orienting bias toward threat was predicted by fractional anisotropy of connections between SC and amygdala in the right but not the left hemisphere.

5.2.2   Asymmetries of Emotional Control Functions in the Ventromedial Prefrontal Cortex

According to most authors (e.g. Drevets and Raichle 1998; Shamay-Tsoory et al. 2003; Pessoa 2010), the ventromedial prefrontal cortex (vmPFC) is especially suitable for integrating diverse cognitive and emotional processes because it receives projections from all sensory modalities and has extensive bidirectional connections with the anterior insular cortex and the amygdala (see Fig. 5.1). Furthermore, the vmPFC is strongly interconnected with brainstem nuclei that are responsible for controlling autonomic and endocrine functions in the service of supporting survival and bodily integrity via homeostasis. Several data suggest that the right vmPFC might have a 'general' role both in the integration between cognition and emotion and in the control of impulsive reactions. Concerning the first issue, Shamay-Tsoory et al. (2003) showed that the right vmPFC regulates the interaction of emotion and cognition in the production of empathic responses. The same authors (Shamay-Tsoory et al. 2005) confirmed that impaired 'affective' theory of mind (TOM) is associated with right vmPFC damage. Furthermore, Zald and Andreotti (2010) showed that damage to the right vmPFC is connected to deficits in detecting irony, sarcasm, and deception. As to the second point, Tranel et al. (2002) demonstrated that patients with lesions of the right vmPFC showed considerably more deficits in social, emotional and decision-making domains than those with left-sided lesions. These subjects also performed defectively and had impaired anticipatory skin conductance responses during the gambling task (see Sect. 3.2 of Chap. 3); this finding, which suggests an important contribution of lesion laterality to impaired decision-making following frontal lobe damage, was confirmed by Clark et al. (2003). These authors administered three neuropsychological measures of decision-making to 46 patients with unilateral PFC lesions and 21 healthy control subjects and found that patients with right frontal lesions preferred the risky decks on the Iowa gambling task, and differed significantly from left frontal and control subjects. Also consistent with the assumption of a greater role of the right vmPFC in the control of impulsive reactions are data reported by Boes et al. (2009), who found a significant correlation between low impulse control and decreased right vmPFC volume in impulsive boys. These data are consistent with the positions of Aron et al. (2004), who, on the basis of human lesion-mapping studies, suggested right lateralisation of control functions subsumed by the the ventral PFC. These authors proposed that a sector of the right inferior frontal cortex (rIFC) implements inhibitory control in both go/no-go and stop signal tasks paradigms via a wider prefrontal-basal ganglia network. The go/no-go test requires that the participant perform an action given certain stimuli (e.g. press a button—Go) and inhibit that action under a different set of stimuli (e.g. do not press that same button—No-Go). The stop signal task is a test of inhibition of prepotent responses. The participant has to respond as quickly as possible to a pre-determined stimulus (the go trial) and to abort any response when a subsequently presented stop signal is displayed. The same authors confirmed these data in a second review performed 10 years later (Aron et al. 2014). A thorough discussion of these positions can be found in Bari and Robbins' (2013) review of the behavioural and neural basis of response control. Data confirming the crucial role of the right frontal cortex in inhibitory control functions have been reported by Wang et al. (2016). In agreement with Bari and Robbins (2013), these authors distinguished two different components of impulsive behaviour (i.e. impulsive choices and impulsive actions) and showed that both of these components are subtended by right frontal structures: the right frontal pole for impulsive choices and the rIFC for impulsive actions. A contrast might be drawn, however, between implicit emotion regulation of the type described by Aron et al. (2004, 2014) and explicit emotion regulation strategies like reappraisal (see Sect. 2.2 of Chap. 2 and Sect. 3.2 of Chap. 3) that rely more heavily on lateral PFC (see Etkin et al. 2015 for a review comparing neural systems for implicit and explicit emotion regulation). Although hemispheric asymmetries in emotion regulation have been studied inadequately, interesting data on this subject were recently obtained by Wyczesany et al. (2018), who used effective connectivity during presentations of positive and negative emotional faces to estimate causal influences between brain regions. These authors documented a strong pattern of connectivity between different frontal areas (i.e. the orbitofrontal and dorsolateral prefrontal cortex/dlPFC) that were mainly lateralised in the right hemisphere for all emotions and a crucial role of the right dlPFC in the top-down regulation of different areas involved in emotional processing. As the dlPFC has long been recognised as critically involved in cognitive control, including cognitive control over emotions (Rilling and Sanfey 2009), these data suggest that the right dlPFC may be crucially involved in explicit emotion regulation strategies.

The lesson that can be drawn from these results (but see Zhang and Zhou 2014 for a different viewpoint) is that they do not support the suggestion, advanced by Gainotti et al. (1993) and reported in Sect. 3.3 of Chap. 3, that the right and left hemispheres may be preferentially involved in different levels of emotional process-ing, with a preferential engagement of the right hemisphere in Leventhal's (1979, 1987) 'schematic level' and a prevalent involvement of the left hemisphere in con-trol functions. These results suggest that right hemisphere dominance could concern both the generation of spontaneous emotions, typical of the 'schematic level', and the control of emotional expression, typical of the 'conceptual level'.

5.2.3   The Contribution of the Right Anterior Insula to the Conscious Experience of Emotion

We have seen in Sect. 3.1 of Chap. 3 that the connectivity of the insular cortex allows for the integration of lower-level homeostatic representations and higher level cognitive functions (Gu et al. 2013) and that, drawing on this connectivity pat-tern, Craig (2009, 2010, 2011) proposed a posterior-to-anterior gradient in the insu-lar cortex. According to this proposal, physical features of interoception should be processed in the posterior insula, whereas the integration of interoception with cog-nitive and motivational information should be processed in the anterior insular cor-tex (AIC). From the experimental point of view, a major contribution of the right AIC to the emotional experience was suggested by Critchley et al. (2004) who mea-sured regional brain activity with fMRI during an interoceptive task in which sub-jects judged the timing of their own heartbeats. These authors observed that neural activity in the right AIC predicted subjects' accuracy in the heartbeat detection task and that greater right AIC volume correlated with increased accuracy in this subjec-tive sense of the inner body and with the corresponding negative emotional experi-ence. According to Critchley et al. (2004), these findings indicate that right AIC supports a representation of visceral responses accessible to awareness and provides a substrate for subjective feeling states.

In another fMRI study, Gray et al. (2007) scanned participants while they judged emotional face stimuli, during both exercise and non-exercise conditions in the con-text of true and false auditory feedback of tonic heart rate. They observed that the perceived emotional intensity/salience of neutral faces was enhanced by false feed-back of increased heart rate and that regional changes in neural activity correspond-ing to this behavioural interaction were observed in right AIC, bilateral mid-insula, and amygdala. Furthermore, BOLD activity within the right AIC and amygdala pre-dicted the corresponding changes in perceived intensity ratings at both the group and the individual level. These findings highlight the importance of the right AIC in guiding second-order 'cognitive' representations of bodily arousal state. Other data supporting the critical role of the right AIC in homeostatic states associated with the different kinds of affectively loaded experiences were reported by Ogino et al. (2007) and by Xue et al. (2010). The former showed that the imagination of pain is associated with increased activity in several brain regions involved in the pain-related neural network (notably the anterior cingulate cortex, the right AIC and the secondary somatosensory cortex region), whereas the latter showed that taking (and especially winning) a gamble was associated with significantly stronger activation in the right AIC. These functions are probably not distinct from the lower-level functions of the insula but rather arise as a consequence of the role of the insula in conveying homeostatic information to consciousness.

From the neuroanatomical point of view, the greater role of the right AIC in emotional functions cound be due to prevalent right lateralisation of the von Economo neurons, which project from the fronto-insular (FI) cortex to the frontal pole, septum and amygdala, relaying information related to autonomic control or awareness of homeostatic representations to these structures. According to Allman et al. (2010), these neurons are indeed about 30% more numerous in the right than in the left FI cortex. A short illustration of the VENs and of their prevalence in the FI cortex is reported in Box 5.1:

Box 5.1 The Von Economo Neurons and Their Prevalence in the Right Fronto-Insular Cortex

The von Economo neurons (VENs) are large, bipolar neurons, located in the fronto-insular (FI) and in the anterior limbic (LA) cortices, that have been observed only in great apes and in humans, but not in other primates.

According to Allman et al. (2010, 2011), the large size and the simple dendritic structure of these projection neurons suggest that they rapidly send basic information related to homeostatic states, autonomic arousal and decision-making from FI and LA cortices to other parts of the brain.

Since the first few months of postnatal life, the number of VENs is greater in the right than in the left hemisphere and in the adult brain neuroimaging studies (e.g. Watkins et al. 2001) have documented a corresponding rightward prevalence of cortical thickness in the FI region containing the VEN neurons.

Allman et al. (2010, 2011) suggested that the rightward asymmetry of the VENs neurons in FI and ACC cortices might be related to an asymmetry of the autonomic nervous system, with a prevalent involvement of its sympathetic section.

A different interpretation of the anatomical substrate of emotional experience in the right and left AIC was proposed by Craig (2005), who assumed that this hemispheric asymmetry was based on an unequal representation of homeostatic activities, resulting from asymmetries in the peripheral autonomic nervous system. This author, who in a first version of his model (Craig 2002, 2003) had acknowledged that the right (non-dominant) anterior insula is integral for the generation of the mental image of one's physical state, which underlies basic emotional states, in a further development of this model drew more strongly on the assumption that forebrain emotional asymmetry might originate from asymmetries in the peripheral autonomic nervous system (see Sect. 4.4 of Chap. 4). He, therefore, maintained that the left AIC was activated predominantly by homeostatic afferents associated with parasympathetic functions, and that the right AI was activated predominantly by homeostatic afferents associated with sympathetic functions (These points will be developed in Sect. 6.2 of Chap. 6).

5.3   Emotional and Behavioural Disorders of Patients with Asymmetrical Forms of Fronto-Temporal Degeneration

Fronto-temporal degeneration (FTD) is a clinically and anatomically heterogeneous disease in which both frontally predominant (frontal variant of FTD [fvFTD]) and temporally predominant (temporal variant of FTD [tvFTD]) subtypes have been described. Typically, tvFTD has been defined by the presence of deficits in language and semantic knowledge, whereas fvFTD has been defined by important behavioural disorders. Several studies (e.g. Bozeat et al. 2000; Snowden et al. 2001; Rosen et al. 2002a; Liu et al. 2004) have, however, shown that both of these anatomo- clinical syndromes may be associated with behavioural disturbances. Furthermore, these and other investigations showed that in the early stages of this disease the atrophies affecting the frontal and (even more) the anterior temporal lobes can be clearly asymmetric (Miller et al. 1993; Rosen et al. 2002b; Thompson et al. 2003; Liu et al. 2004; Seeley et al. 2005) and several studies showed that a relationship exists between emotional and behavioural disorders and side of atrophy.

5.3.1   Relationships Between Emotional and Behavioural Disorders and Left-Right Asymmetry of Atrophy

All investigations that studied the relationships between emotional and behavioural disorders and left-right asymmetry of atrophy showed that these disorders are in the foreground when atrophy prevails in the right frontal or temporal lobes. More precisely, they demonstrated that both behavioural disorders (e.g. Miller et al. 1993; Edwards-Lee et al. 1997; Mychack et al. 2001; Liu et al. 2004; Seeley et al. 2005; Massimo et al. 2009) and emotional comprehension and expression disturbances (e.g. Rosen et al. 2002b, 2006; Irish et al. 2013; Cerami et al. 2015; Kumfor et al. 2016) are mainly observed in FTD patients with a predominance of atrophy in the right orbitofrontal or anterior temporal lobes. In any case, these disorders are related to the dysfunction of these right-sided structures (e.g. Irish et al. 2014; Multani et al. 2017).

Gainotti (2019b) recently reviewed the literature relevant to this issue, in an attempt to evaluate whether the results of investigations conducted on the emotional disorders of patients with right and left FTD mainly support the 'right hemisphere' or the 'valence' hypothesis. Results obtained in 26 investigations were considered as relevant with respect to the 'right hemisphere hypothesis'.

Some of these studies made a comparison between the prevalence of left-sided or right-sided lesions in FTD patients with emotional disorders; in other investigations a significant correlation was observed between emotional or behavioural disorders and atrophy or hypometabolism of left-sided or right-sided brain structures. On the other hand, results obtained in six investigations were considered as relevant with respect to the 'valence hypothesis'. These investigations had made a separate analysis of comprehension or production of positive and negative emotions and compared results obtained on these tasks by FTD patients with a prevalence of left or right lesions or correlated these patterns of emotional disorders with specific left-sided or right-sided structures. Typical examples of investigations relevant to the 'right hemisphere hypothesis' or to the 'valence hypothesis' and conducted with the methodology of 'group comparisons' or 'correlational analysis' between clinical and neuroanatomical data are reported in Box 5.2.

Box 5.2 Examples of Investigations Relevant to the 'Right Hemisphere Hypothesis' or to the 'Valence Hypothesis' in Gainotti's (2019b) Review of the Emotional Disorders of Patients with Asymmetric Forms of Fronto-temporal Degeneration

Hypothesis	Authors	Methodology	Results
Right hemisphere hypothesis	Binney et al. (2016)	**Group comparison.** Compared results obtained by 21 left and 12 right patients with primary progressive aphasia (PPA) on tasks of affect processing	Right greater than left ATL atrophy was associated primarily with early changes in personality and behavioural disturbances such as decreased empathy, blunted affect and deficits in receptive emotional processing
	Kumfor et al. (2016)	**Correlational analysis.** Made a longitudinal psychological and MRI study of 22 left and 9 right SD patients, who were asked to complete a 'face and emotion processing battery' and the 'Cambridge Behavioural inventory'	Even in patients with an initial language presentation, emotional changes reflected right anterior temporal and orbitofrontal cortex degeneration, underscoring the role of these regions in social cognition and behaviour
Valence hypothesis	Irish et al. (2013)	**Group comparison.** Assessed 10 cases of predominantly right and 12 of left semantic dementia on tests of emotion processing (Ekman 60-faces test) and interpersonal functioning (TASIT)	Recognition of happiness was intact in left SD patients, whereas right SD patients displayed profound deficits in the recognition of all basic facial emotions, including happiness
	Werner et al. (2007)	**Correlational analysis.** Assessed in 28 FTD patients emotional reactivity and comprehension of emotional films aimed to elicit fear, happy, and sad emotions and correlated psychological to MRI data	For emotional reactivity, greater happy facial behaviour during the happy film was associated with greater lobar volumes in the right temporal and right frontal lobes, whereas, greater sad facial behaviour during the sad film was associated with greater neuronal volume in the right frontal lobe

Data suggesting that emotional disorders of patients with right FTD concern both negative and positive emotions were also obtained by Hsieh et al. (2012a) in a study in which various groups of patients with degenerative brain diseases completed newly designed emotion word comprehension tasks. Impairment in emotional word knowledge was greatest in SD, compared with bvFTD and AD patients, and word comprehension deficits affected both positive and negative emotions.

The emotional disturbances of patients with predominant atrophy of the right orbitofrontal or anterior temporal lobes can involve a large array of behavioural patterns, ranging from the development of emotion-related behavioural disorders (e.g. Edwards-Lee et al. 1997) to the comprehension of basic emotions (e.g. Rosen et al. 2002b) or the ability to understand and share the feelings of others (e.g. Rankin et al. 2006). For instance, Edwards-Lee et al. (1997) matched the clinical, neuropsychological and neuropsychiatric features of tvFTD patients with predominantly left-sided and right-sided atrophy and found that aphasia was usually the first and most severe clinical abnormality in patients with left-sided atrophy, whereas behavioural disorders (i.e. irritability and impulsiveness) and fixed facial emotional expressions prevailed in patients with right-sided atrophy. The behavioural disturbances of these patients are probably due to the fact that difficulty in perceiving and interpreting the emotions of others can easily give rise to inappropriate behaviour (Liu et al. 2004). On the other hand, Rosen et al. (2002b) examined the comprehension of facial expressions of emotion in nine tvFTD patients and correlated performance on this measure with atrophy of amygdala, ATL and OFC. They showed that emotional comprehension (evaluated with a composite measure of performance on sadness, anger and fear) was correlated with atrophy of the right amygdala. Finally, Rankin et al. (2006) investigated the neuroanatomic basis of empathy in a large sample of patients with different degenerative diseases and found that atrophy in the right temporal pole, the right fusiform gyrus, the right caudate and right subcallosal gyrus correlated significantly with total empathy score. Results consistent with the hypothesis assuming that the emotional disturbances of patients with predominant atrophy of the right anterior temporal lobes may involve a large array of emotion processing were also obtained by Hsieh et al. (2012b) in a correlational study conducted in patients with degenerative brain diseases and designed to evaluate if the same right hemisphere structures implicated in facial emotional recognition are also involved in the processing of emotions in music. Hsieh et al. (2012b) administered to patients with Alzheimer's disease (AD) and Semantic Dementia (SD) tests of emotion recognition in unfamiliar musical tunes and unknown faces and investigated the corresponding neural correlates through voxel-based morphometry. They showed that identification of emotions, regardless of modality, correlated with the degree of atrophy in the right temporal pole, amygdala and insula.

5.3.2   Importance for Emotional Functions of the Right-Sided Connections Between ATL and OFC

If the right anterior temporal and orbitofrontal cortices are critically involved in behavioural and emotional functions, then the connections between ATL and OFC should be particularly strong on the right side. This prediction was confirmed by

Papinutto et al. (2016), who performed a bilateral parcellation of the ATLs based on the degree of connectivity of each voxel with eight ipsilateral target regions. These authors observed two noteworthy interhemispheric differences, because connections between the ATL and the orbitofrontal areas through the uncinate fasciculus (Von Der Heide et al. 2013) were stronger in the right hemisphere, whereas connections between the ATL and the inferior frontal gyrus through the arcuate fasciculus (which has a key role in phonological processing) were greater in the left hemisphere. The relationships between disruption of the right uncinate fasciculus (UF) and behavioural and emotional disturbances were also documented by Craig et al. (2009) and Motzkin et al. (2011) in psychopathic subjects, by Oishi et al. (2015) in stroke patients and by Coad et al. (2017) in healthy subjects. Craig et al. (2009) and Motzkin et al. (2011) found reduced mean fractional anisotropy in the right UF of psychopaths compared with control subjects. Oishi et al. (2015) showed that patients with acute ischemic stroke lesions involving the right UF were significantly more impaired in an emotional empathy task than similar patients without UF lesions, and Coad et al. (2017) showed that interindividual variability in the ability to decode facial emotion expressions in healthy subjects is linked to the right UF microstructure. A schematic representation of the connections between the amygdala and the orbitofrontal areas through the uncinate fasciculus is reported in Fig. 5.2:

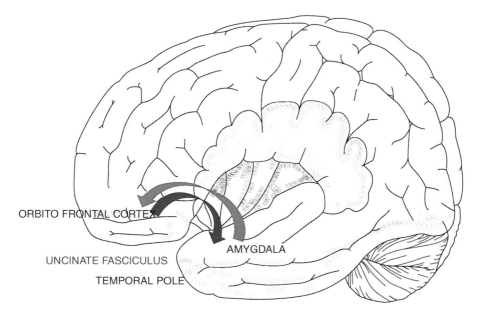

Fig. 5.2 Schematic representation of the connections between the amygdala and the orbitofrontal areas through the uncinate fasciculus, which connects the anterior portions of the temporal lobe with the inferior frontal gyrus, passing inward of the insular cortex. These connections allow an increased ability to decode facial emotion expressions and a 'top-down' modulation of intense emotional responses

5.3.3   Investigations Which Have Checked Specific Models of Emotional Lateralisation

Investigations conducted on patients with asymmetrical forms of fronto-temporal degeneration also tried to directly check some of the models advanced to explain the relationships between emotions and brain laterality. For example Guo et al. (2016), drawing on the controversy raised by Craig (2002, 2005, 2009) concerning the hemispheric lateralisation of emotions along the parasympathetic-sympathetic axis, showed that in bvFTD a reduced baseline cardiac vagal tone correlates with left-lateralised fronto-insular and cingulate cortex deficits and with reduced agreeableness. They therefore suggested that the left-sided parts of the 'salience network' (a distributed neural system, primarily composed of the AIC and of the ACC, that maintains homeostasis by regulating autonomic nervous system activity and social-emotional functions) are critical for maintaining an adaptive level of baseline parasympathetic outflow.

The main characteristics of the 'salience network' are reported in Box 5.3:

Box 5.3 Main Features of the 'Salience Network'

According to an emerging paradigm in cognitive neuroscience, cognitive tasks are not performed by individual brain regions working in isolation, but by multiple, distinct and interacting 'large scale brain networks'. These networks consist of discrete brain regions that are said to be functionally connected, due to their tightly coupled activity, measured as long-range syncronization of EEG, MEG or other dynamic brain signals (Bressler 1995; Bressler and Menon 2010).

One of these large scale brain network is the salience network (SN), which includes the AIC and the dorsal anterior cingulate cortex (dACC). The SN is deemed to be involved in detecting and filtering salient stimuli (enabling the organism to focus its limited resources on the most pertinent sensory data) and recruiting the corresponding relevant functional networks (Menon and Uddin 2010).

According to Seeley et al. (2007, 2008), the afferent interoceptive component of the SN, which processes the major afferent input streams regarding homeostatically relevant internal stimuli, should be based on the AIC, whereas its efferent component should be centred on the dACC. This component should mobilise viscero-autonomic responses to salience and recruit executive and task control resources to guide behaviour.

In further investigations, Sturm et al. (2018a, b) examined how within-salience network connectivity relates to individual differences in human baseline parasympathetic and sympathetic nervous activity and examined whether resting parasympathetic deficits in bvFTD were related to diminished prosocialbehaviour. In the first study, Sturm et al. (2018a) used voxel-based morphometry to determine whether salience network atrophy was associated with lower baseline respiratory sinus arrhythmia (a parasympathetic measure) and skin conductance level (a sympathetic measure) in bvFTD. They showed that lower baseline respiratory sinus arrhythmia was associated with smaller volume in the left ventral anterior insula (vAI) and weaker connectivity between the bilateral vAI and the bilateral anterior cingulate cortex (ACC), whereas lower baseline skin conductance level was associated with smaller volume in the inferior temporal gyrus, dorsal mid-insula and hypothalamus and weaker connectivity between the bilateral ACC and the right hypothalamus/amygdala. They concluded that baseline parasympathetic and sympathetic tone depends on the integrity of lateralised salience network hubs (left vAI for parasympathetic and right hypothalamus/amygdala for sympathetic) and highly calibrated ipsilateral and contralateral network connections. In the second study, Sturm et al. (2018b) showed that left-lateralised fronto-insular atrophy is associated not only with lower respiratory sinus arrhythmia but also with lower consolation and greater disengagement, suggesting that left-lateralised salience network atrophy reduces patients' resting parasympathetic activity and motivation to help others in need. One problem raised by these findings concerns their relevance (or, more in general, their implications) with respect to the general models of emotion lateralisation discussed in previous chapters of this volume. They, therefore, will be taken again into account in Chap. 6 and in the concluding remarks of this volume.

References

Adolphs R, Damasio H, Tranel D, Cooper G, Damasio AR. A role for somatosensory cortices in the visual recognition of emotion as revealed by three-dimensional lesion mapping. J Neurosci. 2000;20:2683–90.

Allman JM, Tetreault NA, Hakeem AY, Manaye KF, Semendeferi K, Erwin JM, et al. The von Economo neurons in frontoinsular and anterior cingulate cortex in great apes and humans. Brain Struct Funct. 2010;214:495–517.

Allman JM, Tetreault NA, Hakeem AY, Manaye KF, Semendeferi K, Erwin JM, et al. The von Economo neurons in frontoinsular and anterior cingulate cortex. Ann N Y Acad Sci. 2011;1225:59–71.

Aron AR, Robbins TW, Poldrack RA. Inhibition and the right inferior frontal cortex. Trends Cogn Sci. 2004;8:170–7.

Aron AR, Robbins TW, Poldrack RA. Inhibition and the right inferior frontal cortex: one decade on. Trends Cogn Sci. 2014;18:177–85.

Balconi M, Lucchiari C. Consciousness and arousal effects on emotional face processing as revealed by brain oscillations. A gamma band analysis. Int J Psychophysiol. 2008;67:41–6.

Bari A, Robbins TW. Inhibition and impulsivity: behavioral and neural basis of response control. Prog Neurobiol. 2013;108:44–79.

Bertini C, Cecere R, Làdavas E. Unseen fearful faces facilitate visual discrimination in the intact field. Neuropsychologia. 2019;128:58–64.

Binney RJ, Henry ML, Babiak M, Pressman PS, Santos-Santos MA, Narvid J, et al. Reading words and other people: a comparison of exception word, familiar face and affect processing in the left and right temporal variants of primary progressive aphasia. Cortex. 2016;82:147–63.

Boes AD, Bechara A, Tranel D, Anderson SW, Richman L, Nopoulos P. Right ventromedial prefrontal cortex: a neuroanatomical correlate of impulse control in boys. Soc Cogn Affect Neurosc. 2009;4:1–9.

Bozeat S, Gregory CA, Lambon Ralph MA, Hodges JR. Which neuropsychiatric and behavioural features distinguish frontal and temporal variants of frontotemporal dementia from Alzheimer's disease? J Neurol Neurosurg Psychiatry. 2000;69:178–86.

Bressler SL. Large-scale cortical networks and cognition. Brain Res Brain Res Rev. 1995;20:288–304.

Bressler SL, Menon V. Large-scale brain networks in cognition: emerging methods and principles. Trends Cogn Sci. 2010;14:277–90.

Burra N, Hervais-Adelman A, Celeghin A, de Gelder B, Pegna AJ. 2017. Affective blindsight relies on low spatial frequencies. Neuropsychologia 2019;128:44–49.

Cecere R, Bertini C, Maier ME, Làdavas E. Unseen fearful faces influence face encoding: evidence from ERPs in hemianopic patients. J Cogn Neurosci. 2014;26:2564–77.

Cerami C, Dodich A, Iannaccone S, Marcone A, Lettieri G, Crespi C, et al. Right limbic FDG-PET hypometabolism correlates with emotion recognition and attribution in probable behavioral variant of frontotemporal dementia patients. PLoS One. 2015;10:e0141672. https://doi.org/10.1371/journal.pone.0141672. eCollection 2015.

Clark L, Manes F, Nagui A, Sahakian BJ, Robbins TW. The contributions of lesion laterality and lesion volume to decision-making impairment following frontal lobe damage. Neuropsychologia. 2003;41:1474–83.

Coad BM, Postans M, Hodgetts CJ, Muhlert N, Graham KS, Lawrence AD. Structural connections support emotional connections: uncinate fasciculus microstructure is related to the ability to decode facial emotion expressions. Neuropsychologia. 2017. pii: S0028–3932(17)30420–7. https://doi.org/10.1016/j.neuropsychologia.2017.11.006. [Epub ahead of print].

Craig AD. A new view of pain as a homeostatic emotion. Trends Neurosci. 2002;26:303–7.

Craig AD. Interoception: the sense of the physiological condition of the body. Curr Opin Neurobiol. 2003;13:500–5.

Craig AD. Forebrain emotional asymmetry: a neuroanatomical basis? Trends Cogn Sci. 2005;9:566–71.

Craig AD. How do you feel—now? The anterior insula and human awareness. Nat Rev Neurosci. 2009;10:59–70.

Craig AD. The sentient self. Brain Struct Funct. 2010;214:563–77.

Craig AD. Significance of the insula for the evolution of human awareness of feelings from the body. Ann N Y Acad Sci. 2011;1225:72–82.

Craig MC, Catani M, Deeley Q, Latham R, Daly E, Kanaan R, et al. Altered connections on the road to psychopathy. Mol Psychiatry. 2009;946-53(907):14.

Critchley HD, Wiens S, Rotshtein P, Ohman A, Dolan RJ. Neural systems supporting interoceptive awareness. Nat Neurosci. 2004;7:189–95.

Drevets WC, Raichle ME. Reciprocal suppression of regional cerebral blood flow during emotional versus higher cognitive processes: implications for interactions between cognition and emotion. Cogn Emot. 1998;12:353–85.

Edwards-Lee T, Miller BL, Benson DF, Cummings JL, Russell GL, Boone K, et al. The temporal variant of frontotemporal dementia. Brain. 1997;120:1027–40.

Etkin A, Büchel C, Gross JJ. The neural bases of emotion regulation. Nat Rev Neurosci. 2015;16:693–700.

Eugène F, Lévesque J, Mensour B, Leroux JM, Beaudoin G, Bourgouin P, et al. The impact of individual differences on the neural circuitry underlying sadness. NeuroImage. 2003;19:354–64.

Fusar-Poli P, Piacentino A, Carletti F, Allen P, Landi P, Abbamonte M, et al. Laterality effect on emotional faces processing: ALE meta-analysis of evidence. Neurosci Lett. 2009;452:262–7.

Gainotti G. Unconscious processing of emotions and the right hemisphere. Neuropsychologia. 2012;50:205–18.

Gainotti G. Emotions and the right hemisphere: can new data clarify old models? Neuroscientist. 2019a;25:258–70.

Gainotti G. The role of the right hemisphere in emotional and behavioral disorders of patients with frontotemporal lobar degeneration: an updated review. Front Aging Neurosci. 2019b; https://doi.org/10.3389/fnagi.2019.00055.

Gainotti G, Caltagirone C, Zoccolotti P. Left/right and cortical subcortical dichotomies in the neuropsychological study of human emotions. Cogn Emot. 1993;7:71–93.

Gläscher J, Adolphs R. Processing of the arousal of subliminal and supraliminal emotional stimuli by the human amygdala. J Neurosci. 2003;23:10274–82.

Gray MA, Harrison NA, Wiens S, Critchley HD. Modulation of emotional appraisal by false physiological feedback during fMRI. PLoS One. 2007;2:e546. https://doi.org/10.1371/journal.pone.0000546.

Gu X, Hof PR, Friston KJ, Fan J. Anterior insular cortex and emotional awareness. J Comp Neurol. 2013;521:3371–88.

Guo CC, Sturm VE, Zhou J, Gennatas ED, Trujillo AJ, Hua AY, et al. Dominant hemisphere lateralization of cortical parasympathetic control as revealed by frontotemporal dementia. Proc Natl Acad Sci U S A. 2016;113:E2430–9.

Hsieh S, Foxe D, Leslie F, Savage S, Piguet O, Hodges JR. Grief and joy: emotion word comprehension in the dementias. Neuropsychology. 2012a;26:624–30.

Hsieh S, Hornberger M, Piguet O, Hodges JR. Brain correlates of musical and facial emotion recognition: evidence from the dementias. Neuropsychologia. 2012b;50:1814–22.

Hung Y, Smith ML, Bayl DJ, Mills T, Cheyne D, Taylor MJ. Unattended emotional faces elicit early lateralized amygdala frontal and fusiform activations. NeuroImage. 2010;50:727–33.

Irish M, Kumfor F, Hodges JR, Piguet O. A tale of two hemispheres: contrasting socioemotional dysfunction in right- versus left-lateralised semantic dementia. Dement Neuropsychol. 2013;7:88–95.

Irish M, Hodges JR, Piguet O. Right anterior temporal lobe dysfunction underlies theory of mind impairments in semantic dementia. Brain. 2014;137:1241–53.

Koller K, Rafal RD, Platt A, Mitchell ND. Orienting toward threat: contributions of a subcortical pathway transmitting retinal afferents to the amygdala via the superior colliculus and pulvinar. Neuropsychologia. 2019;128:78–86.

Kumfor F, Landin-Romero R, Devenney E, Hutchings R, Grasso R, Hodges JR, et al. On the right side? A longitudinal study of left- versus right-lateralized semantic dementia. Brain. 2016;139:986–98.

LeDoux J. The emotional brain. New York: Simon and Schuster; 1996.

LeDoux JE, Sakaguchi A, Iwata J, Reis DJ. Interruption of projections from the medial geniculate body to an archi-neostriatal field disrupts the classical conditioning of emotional responses to acoustic stimuli in the rat. Neuroscience. 1986;17:615–27.

Levenson RW, Sturm VE, Haase CM. Emotional and behavioral symptoms in neurodegenerative disease: a model for studying the neural bases of psychopathology. Annu Rev Clin Psychol. 2014;10:581–606.

Leventhal H. A perceptual- motor processing model of emotion. In: Pliner P, Blankestein K, Spiegel IM, editors. Perception of emotion in self and others, vol. 5. New York: Plenum; 1979. p. 1–46.

Leventhal H. A perceptual motor theory of emotion. In: Berkowitz L, editor. Advances in experimental social psychology, vol. 17. New York: Academic Press; 1987. p. 117–82.

Lindquist KA, Satpute AB, Wager TD, Weber J, Barrett LF. The brain basis of positive and negative affect: evidence from a meta-analysis of the human neuroimaging literature. Cereb Cortex. 2016;26:1910–22.

Liu W, Miller BL, Kramer JH, Rankin K, Wyss-Coray C, Gearhart R, et al. Behavioral disorders in the frontal and temporal variants of frontotemporal dementia. Neurology. 2004;62:742–8.

Luo Q, Mitchell D, Cheng X, Mondillo K, Mccaffrey D, Holroyd T, et al. Visual awareness, emotion, and gamma band synchronization. Cereb Cortex. 2009;19:1896–904.

Massimo L, Powers C, Moore P, Vesely L, Avants B, Gee J, et al. Neuroanatomy of apathy and disinhibition in frontotemporal lobar degeneration. Dement Geriatr Cogn Disord. 2009;27:96–104.

Meletti S, Benuzzi F, Rubboli G, Cantalupo G, Stanzani Maserati M, Nichelli P, et al. Impaired facial emotion recognition in early-onset right mesial temporal lobe epilepsy. Neurology. 2003;60:426–31.

Meletti S, Benuzzi F, Cantalupo G, Rubboli G, Tassinari CA, Nichelli P. Facial emotion recognition impairment in chronic temporal lobe epilepsy. Epilepsia. 2009;50:1547–59.

Menon V, Uddin LQ. Saliency, switching, attention and control: a network model of insula function. Brain Struct Funct. 2010;214:655–67.

Miller BL, Chang L, Mena I, Boone K, Lesser IM. Progressive right frontotemporal degeneration: clinical, neuropsychological and SPECT characteristics. Dementia. 1993;4:204–13.

Morris JS, Ohman A, Dolan RJ, Rowland D, Young AW, Calder AJ, et al. Conscious and unconscious emotional learning in the human amygdala. Nature. 1998;393:467–70.

Morris JS, Ohman A, Dolan RJ. A subcortical pathway to the right amygdala mediating 'unseen' fear. PNAS. 1999;96:1680–5.

Motzkin JC, Newman JP, Kiehl KA, Koenigs M. Reduced prefrontal connectivity in psychopathy. J Neurosci. 2011;31:17348–57.

Multani N, Galantucci S, Wilson SM, Shany-Ur T, Poorzand P, Growdon ME, et al. Emotion detection deficits and changes in personality traits linked to loss of white matter integrity in primary progressive aphasia. NeuroImage: Clin. 2017;16:447–54.

Mychack P, Kramer JH, Boone KB, Miller BL. The influence of right frontotemporal dysfunction on social behavior in frontotemporal dementia. Neurology. 2001;56:suppl 4:11–5.

Noesselt T, Driver J, Heinze HJ, Dolan R. Asymmetrical activation in the human brain during processing of fearful faces. Curr Biol. 2005;15:424–9.

Nomura M, Ohira H, Haneda K, Iidaka T, Sadato N, Okada T, Yonekura Y. Functional association of the amygdala and ventral prefrontal cortex during cognitive evaluation of facial expressions primed by masked angry faces: an event related fMRI study. NeuroImage. 2004;21:352–63.

Ogino Y, Nemoto H, Inui K, Saito S, Kakigi R, Goto F. Inner experience of pain: imagination of pain while viewing images showing painful events forms subjective pain representation in human brain. Cereb Cortex. 2007;17:1139–46.

Oishi K, Faria AV, Hsu J, Tippett D, Mori S, Hillis AE. Critical role of the right uncinate fasciculus in emotional empathy. Ann Neurol. 2015;77:68–74.

Papez JW. A proposed mechanism of emotion. Arch Neurol Psychiatr. 1937;79:217–24.

Papinutto N, Galantucci S, Mandelli ML, Gesierich B, Jovicich J, Caverzasi E, et al. Structural connectivity of the human anterior temporal lobe: a diffusion magnetic resonance imaging study. Hum Brain Mapp. 2016;37:2210–22.

Pegna AJ, Landis T, Khateb A. Electrophysiological evidence for early non-conscious processing of fearful facial expressions. Int J Psychophysiol. 2008;70:127–36.

Pessoa L. Emergent processes in cognitive-emotional interactions. Dialogues Clin Neurosci. 2010;12:433–48.

Rilling JK, Sanfey AG. Social interaction in encyclopedia of neuroscience, squire L (ed.) Vol. 9. London: Academic; 2009. p. 41–8.

Rosen HJ, Gorno-Tempini ML, Goldman WP, Schuff N, Weiner M, et al. Patterns of brain atrophy in frontotemporal dementia and semantic dementia. Neurology. 2002a;58:198–208.

Rosen HJ, Perry RJ, Murphy J, Kramer JH, Mychack P, Schuff N, et al. Emotion comprehension in the temporal variant of frontotemporal dementia. Brain. 2002b;125:2286–95.

Rosen HJ, Wilson MR, Schauer GF, Allison S, Gorno-Tempini ML, Pace-Savitsky C, et al. Neuroanatomical correlates of impaired recognition of emotion in dementia. Neuropsychologia. 2006;44:365–73.

Schepman A, Rodway P, Cornmell L, Smith B, de Sa SL, Borwick C, et al. Right-ear precedence and vocal emotion contagion: the role of the left hemisphere. Laterality. 2018;23:290–317.

Seeley WW, Bauer AM, Miller BL, Gorno-Tempini ML, Kramer JH, Weiner M, et al. The natural history of temporal variant frontotemporal dementia. Neurology. 2005;64:1384–90.

Seeley WW, Menon V, Schatzberg AF, Keller J, Glover GH, Kenna H, et al. Dissociable intrinsic connectivity networks for salience processing and executive control. J Neurosci. 2007;27:2349–56.

Seeley WW, Crawford R, Rascovsky K, Kramer JH, Weiner M, Miller BL, et al. Frontal paralimbic network atrophy in very mild behavioral variant frontotemporal dementia. Arch Neurol. 2008;65:249–55.

Shamay-Tsoory SG, Tomer R, Berger BD, Aharon-Peretz J. Characterization of empathy deficits following prefrontal brain damage: the role of the right ventromedial prefrontal cortex. J Cogn Neurosci. 2003;15:324–37.

Shamay-Tsoory SG, Tomer R, Berger BD, Goldsher D, Aharon-Peretz J. Impaired "affective theory of mind" is associated with right ventromedial prefrontal damage. Cogn Behav Neurol. 2005;18:55–67.

Snowden JS, Bathgate D, Varma A, Blackshaw A, Gibbons ZC, Neary D. Distinct behavioural profiles in frontotemporal dementia and semantic dementia. J Neurol Neurosurg Psychiatry. 2001;70:323–32.

Sturm VE, Sible IJ, Datta S, Hua AY, Perry DC, Kramer JH et al. Resting parasympathetic dysfunction predicts prosocial helping deficits in behavioural variant frontotemporal dementia. Cortex 2018a;109:141–55.

Sturm VE, Brown JA, Hua AY, Lwi SJ, Zhou J, Kurth F et al. Network architecture underlying basal autonomic outflow: evidence from frontotemporal dementia. J Neurosci. 2818b;38:8943–55.

Thompson SA, Patterson K, Hodges JR. Left/right asymmetry of atrophy in semantic dementia: behavioral-cognitive implications. Neurology. 2003;61:1196–203.

Tippett DC, Godin BR, Oishi K, Oishi K, Davis C, Gomez Y, et al. Impaired recognition of emotional faces after stroke involving right amygdala or insula. Semin Speech Lang. 2018;39:87–100.

Tranel D, Bechara A, Denburg NL. Asymmetric functional roles of right and left ventromedial prefrontal cortices in social conduct, decision-making, and emotional processing. Cortex. 2002;38:589–612.

Von Der Heide RJ, Skipper LM, Klobusicky E, Olson IR. Dissecting the uncinate fasciculus: disorders, controversies and a hypothesis. Brain. 2013;136:1692–707.

Wager TD, Phan KL, Liberzon I, Taylor SF. Valence, gender, and lateralization of functional brain anatomy in emotion: a meta-analysis of findings from neuroimaging. NeuroImage. 2003;19:513–31.

Wang Q, Chen C, Cai Y, Li S, Zhao X, Zheng L, et al. Dissociated neural substrates underlying impulsive choice and impulsive action. NeuroImage. 2016;134:540–9.

Watkins KE, Paus T, Lerch JP, Zijdenbos A, Collins DL, Neelin P, et al. Structural asymmetries in the human brain: a voxel-based statistical analysis of 142 MRI scans. Cereb Cortex. 2001;11:868–77.

Werner KH, Roberts NA, Rosen HJ, Dean DL, Kramer JH, Weiner MW, et al. Emotional reactivity and emotion recognition in frontotemporal lobar degeneration. Neurology. 2007;69:148–55.

Williams MA, Mattingley JB. Unconscious perception of non-threatening facial emotion in parietal extinction. Exp Brain Res. 2004;154:403–6.

Williams LM, Das P, Liddell BJ, Kemp AH, Rennie CJ, Gordon E. Mode of functional connectivity in amygdala pathways dissociates level of awareness for signals of fear. J Neurosci. 2006;26:9264–71.

Wright CI, Martis B, Schwartz CE, Shin LM, Fischer H, McMullin K, et al. Novelty responses and differential effects of order in the amygdala, substantia innominata, and inferior temporal cortex. NeuroImage. 2003;18:660–9.

Wyczesany M, Capotosto P, Zappasodi F, Prete G. Hemispheric asymmetries and emotions: evidence from effective connectivity. Neuropsychologia. 2018;121:98–105.

Xue G, Lu Z, Levin IP, Bechara A. The impact of prior risk experiences on subsequent risky decision-making: the role of the insula. NeuroImage. 2010;50:709–16.

Zald DH, Andreotti C. Neuropsychological assessment of the orbital and ventromedial prefrontal cortex. Neuropsychologia. 2010;48:3377–91.

Zhang J, Zhou R. Individual differences in automatic emotion regulation affect the asymmetry of the LPP component. PLoS One. 2014;9(2):e88261. https://doi.org/10.1371/journal.pone.0088261. eCollection 2014.

General Neurobiological Models Advanced to Explain Results Obtained Following these New Lines of Research

6

Contents

6.1 Controversies About Results Obtained in Patients with Asymmetrical Forms of Fronto-Temporal Degeneration

In Chap. 5, we saw that data obtained studying laterality effects in brain structures that have specific roles in emotional functions (Sect. 5.2) and those obtained by investigating behavioural and emotional disorders of patients with asymmetrical forms of fronto-temporal degeneration (Sect. 5.3) have provided an updated, but still controversial, version of old models of emotional lateralisation. Data obtained studying laterality effects in brain structures that have a critical role in different emotional functions (such as the amygdala, vmPFC and AIC) tend to support the 'right hemisphere hypothesis' more than the 'valence hypothesis' and to provide a more articulated view of these laterality effects than the traditional versions, which viewed the right and the left hemisphere as unitary functional units. These studies have, in fact, shown that the RH dominance for emotions concerns all components

© Springer Nature Switzerland AG 2020   73
G. Gainotti, *Emotions and the Right Side of the Brain*,
https://doi.org/10.1007/978-3-030-34090-2_6

of the emotional network, from the unconscious processing of emotional information in the right amygdala, to the conscious experience of emotions in the right AIC, to the representation and 'topdown' modulation of emotional responses in the right inferior frontal cortex. On the other hand, some investigations conducted on patients with asymmetrical forms of fronto-temporal degeneration and reported in the last section of Chap. 5 suggest that the model of a different lateralisation of positive and negative emotions could be reformulated in terms of sympathetic vs parasympathetic functions. Guo et al. (2016) and Sturm et al. (2018a, b) suggested that atrophy of the left-lateralised salience network reduces prosocial activities supported by parasympathetic functions. However, this suggestion raises the problem of the relationships between 'prosocial activities' and basic positive emotions, stressed by the 'valence hypothesis', or 'approach tendencies', underlined by the 'approach/withdrawal' motivational hypothesis. At a first glance, no link should exist between prosocial activities and 'positive emotions', whereas some overlap might be found between these activities and 'approach tendencies'. However, a closer look shows that even this analogy is not without problems. In the discussion concerning the relations between positive emotions and approach tendencies (see Sect. 4.3 of Chap. 4), we saw that the only basic emotion in which a dissociation can be found between valence and motivational tendencies is anger because it is a negative emotion that evokes an approach motivation (Russell and Barrett 1999). Several authors (e.g. Harmon-Jones and Allen 1998; Harmon-Jones and Sigelman 2001; Hewig et al. 2004) showed, however, that anger is associated with relative left-prefrontal activity, but this left lateralisation of hanger is inconsistent with the similar lateralisation of prosocial activities and with the fact that anger is supported by sympathetic activities (Cannon 1927; Frijda 1986, 1987). It could be objected that the brain structures which subserve the basic emotions (on which the classical models of emotion lateralisation were constructed) are probably different from the brain structures which are damaged in patients with asymmetrical forms of fronto-temporal degeneration and which subserve complex social emotions on which Sturm et al.'s (2018a, b) results were based. Two main counterarguments, based respectively on data gathered by Gainotti (2019b) in his review of the relationships between emotional disorders and laterality of atrophy in patients with asymmetric forms of fronto-temporal degeneration and on the study of laterality effects in brain structures that have a different role in emotional functions (surveyed in Sect. 5.2 of Chap. 5) can, however, be raised to counter this objection. The first is that data gathered by Gainotti (2019b) concerned basic more than complex social emotions, but strongly supported the 'right hemisphere hypothesis'. The second is that the unconscious processing of emotions in the right amygdala, the greater role of the right vmPFC in the control of impulsive reactions and the greater contribution of the right AIC to the conscious experience of emotion concerned basic more than complex/social emotions. Furthermore, data attenuating the general value of the Sturm et al.'s (2018a, b) statements that atrophy of the left-lateralised salience network (SN) reduces prosocial activities supported by parasympathetic functions have been recently reported by Toller et al. (2018, 2019). Toller et al. (2018) examined how individual differences in SN connectivity are reflected in overt social behaviour in healthy individuals and patients with degenerative brain diseases and showed that individual differences in

SN functional connectivity are directly related to individual differences in observed socioemotional sensitivity. In partial disagreement with the Sturm et al.'s (2018a, b) data, they showed, however, that socioemotional sensitivity among healthy individuals was mainly predicted by high functional connectivity in the SN, between the right (rather than left) AIC and 'interoceptive' subcortical (dorsomedial thalamus, hypothalamus, amygdala, PAG) nodes. On the other hand, trying to analyse individual differences in the loss of interpersonal warmth in FTD syndromes, Toller et al. (2019) observed that a drastic loss of interpersonal warmth is seen only in a subset of patients with either behavioural variant or semantic variant FTD, thus questioning the strength of the link, outlined by Sturm et al. (2018a, b) between atrophy of the left-lateralised salience network and reduction of prosocial activities in this disease. Although these comments suggest great caution before drawing any conclusion on this controversial issues, we can say that two general neurobiological models have recently been advanced by Gainotti (2005, 2007a, 2012, 2019a) and by Craig (2005, 2009, 2010, 2011) to account for results obtained following these new lines of research. These models are based on divergent evolutionary explanations and on the revival of previous models that assume, respectively: (a) a general right hemisphere dominance for emotional functions and (b) different specialisation of the two sides of the brain for different aspects of emotions.

6.2   A Model Based on the General Assumption of a More Primitive Functional Organisation of the Nonverbal Right Hemisphere

Gainotti's (2005, 2007a, 2012, 2019a) model draws on the general account that many authors (e.g. Ekman 1984; Frijda 1986; LeDoux 1996; Panksepp 2010) have given of emotion, i.e. that it is a phylogenetically older, emergency adaptive system. This model assumes that these features of the emotional system may be consonant with the nonverbal functional organisation of the right hemisphere, considered more primitive than the verbally shaped organisation of the left hemisphere. According to this model, a high degree of emotional processing, reliance on sensorimotor functions, unawareness and automaticity should characterise the nonverbal functional organisation of the right hemisphere, whereas a prevalence of verbal cognitive processing, consciousness and intentionality should characterise the left hemisphere's functional organisation. Gainotti (2019a) maintained that this position is supported by two main sources of evidence: (a) the different format of representations subsumed by the right and left anterior temporal lobes and (b) the distinction between automatic and controlled processing of information carried out by the right and left hemispheres. The first point is based on evidence showing that both conceptual and familiar people representations are stored in a sensory, embodied, nonverbal format in the right ATL and in a verbal modality in the left ATL (Snowden et al. 2004; Gainotti 2007b, 2012, 2015; Woollams and Patterson 2018). The second point is based on the following sources of evidence: (A) data obtained by Morris et al. (1998, 1999) and extensively discussed in Sect. 5.2.1 of this volume, which show that the right amygdala plays a major role in automatic/unconscious and the left in controlled/conscious forms of emotional learning; (B) data reviewed by Vuilleumier et al. (2003) and by Gainotti (2007a, 2011), which show that familiarity feelings, involved in the recognition of familiar people, are automatically generated by right hemisphere structures; (C) data, reviewed by Gainotti (1996) and by Bartolomeo and Chokron (2002) which show that the syndrome of unilateral spatial neglect, which is characteristic of right hemisphere lesions, is due to disruption of the automatic (rather than of the controlled) components of the spatial orienting of attention.

6.3   A Model Assuming That Different Homeostatic Models of Awareness Originate from Asymmetries in the Peripheral Autonomic Nervous System

According to the homeostatic model of awareness proposed by Craig (2005, 2009, 2010, 2011), hemispheric asymmetries for emotional experience might be based on an unequal representation of homeostatic activities, resulting from asymmetries in the peripheral autonomic nervous system. Due to these peripheral asymmetries, parasympathetic activities should be mainly represented in the left hemisphere, and sympathetic activities should be mainly represented in the right hemisphere. Homeostatic activities in the right side of the forebrain might, therefore, be associated with energy expenditure, sympathetic activity, arousal, withdrawal (aversive) behaviour and individual-oriented (survival) emotions, whereas the homologous activities in the left side might be associated with energy nourishment, parasympathetic activity, relaxation, approach (appetitive) behaviour and group-oriented (affiliative) emotions. To support this model, which is based on the coordinated opponency of the sympathetic and parasympathetic components of the autonomic nervous system, Craig (2005, 2009, 2010) mainly relied on data from the neuropsychological and psychophysiological literature (reviewed in previous sections of the present volume) which suggested that the left and right halves of the forebrain might be differentially associated with positive and negative affect (Sect. 4.3), approach and withdrawal tendencies (Sect. 4.3) and sympathetic and parasympathetic activities (Sect. 4.4).

6.4   Relationships Between the Lateralisation of Autonomic and of Emotional Activities

In Sect. 4.4 of the present volume, we saw that data obtained studying the lateralisation of autonomic activities in normal subjects and in patients with unilateral brain damage consistently suggest an asymmetrical representation of the autonomic functions in the human brain. We also saw, however, that this claim does not equally concern both components of the autonomic system. The contrast between the conclusions reached on this subject by Wittling and coworkers (Wittling 1995; Wittling et al. 1998) and by Spence et al. (1996) in their studies of the psychophysiological correlates of the selective emotional stimulation of the right and left hemispheres in normal subjects (see Sect. 4.4 of Chap. 4), clearly illustrates this point. More in general, all authors agree on the leading role of the right hemisphere in the modulation of sympathetic activities; however, the lateralisation of parasympathetic functions is more controversial. It is also interesting to note that a strong similarity exists between models advanced to explain the hemispheric asymmetries for autonomic functions and models previously advanced to explain the lateralisation of emotional functions in the human brain. In both cases one model assumes a general superiority of the right hemisphere for emotions and autonomic functions, whereas the other model assumes a different hemispheric specialisation (for positive vs negative emotions or for approach vs avoidance tendencies, or for sympathetic vs parasympathetic components of the vegetative system). Furthermore, just as a right hemisphere superiority for negative emotions is almost universally acknowledged and a left hemisphere prevalence for positive emotions is supported only by some studies, a right hemisphere dominance for sympathetic functions is much more supported than a left hemisphere prevalence for parasympathetic activities. In fact, almost all data summarised in Sect. 4.4 of Chap. 4 suggest right hemisphere dominance for sympathetic activities, whereas data concerning hemispheric asymmetries for parasympathetic functions are more controversial. On the one hand, insular stimulation data, obtained by Oppenheimer et al. (1992), suggest that sympathetic cardiovascular regulation is mainly a right insular function, whereas parasympathetic cardiac neural regulation predominates in the left insula. In line with these results, Wittling et al. (1998) reported data which strongly support the hypothesis of left hemisphere superiority for parasympathetic activities. More recently, Guo et al. (2016) and Sturm et al. (2018a, b) obtained results that reinforce this hypothesis. They showed that in patients with asymmetrical forms of FTD baseline parasympathetic tone depends on the integrity of the left ventral anterior insula. On the other hand, data obtained by Zoccolotti et al. (1986), Yokoyama et al. (1987), Caltagirone et al. (1989), Làdavas et al. (1993), Barron et al. (1994), Naver et al. (1996), Andersson and Finset (1998), Tokgozoglu et al. (1999), Spence et al. (1996) and Colivicchi et al. (2004) seem to point to a right hemisphere dominance for parasympathetic as well as for sympathetic activities. Therefore, it seems safe to conclude that a difference exists between the strong right lateralisation of sympathetic activities and the weak left lateralisation of parasympathetic activities.

6.5 Asymmetrical Representation of Emotions and of Central Representation of Homeostatic Activities in Nonhuman Primates

In his discussion of the model proposing that the lateralisation of emotional functions may be due to an asymmetrical central representation of homeostatic activities, Craig (2005) acknowledges that in the forebrain of nonhuman primates there is little evidence of lateralised homeostatic afferent processing (Hanamori et al. 1998; Shi and Cassell 1998). On the contrary, in humans there is strong evidence for functional lateralisation of homeostatic afferent pathways. Afferents from parasympathetic activities are, in fact, lateralised to the left in the nucleus of the solitary tract of the brainstem (Craig 2002), whereas inputs from sympathetic activities are lateralised to the right in lamina I of the spinal dorsal horn (Craig et al. 2000). From these structures, these complementary vegetative systems activate higher-order homeostatic re-representations in more anterior portions of the (left and right) human insula. Thus, the left anterior insula could be predominantly activated by homeostatic afferents associated with the parasympathetic system and the right AI by homeostatic afferents associated with sympathetic functions. However if emotional functions are due to an asymmetrical central representation of homeostatic activities, and if the homeostatic afferents are lateralised only in humans, it should be predicted that a lateralisation of emotional functions should only be observed in humans. This prediction is nevertheless disconfirmed by several empirical investigations, which have documented the existence of continuities in emotion lateralisation in human and nonhuman primates. For instance, Wallez and Vauclair (2013), using the chimeric face technique (see Chap. 4, Sect. 4.3.1) showed that the left-composite chimeric faces of baboons and humans are judged to be emotionally stronger than the right ones for the emotional behaviours, but not for the neutral non emotional category. Similar results were obtained by Lindell (2013), who reviewed research examining the patterns of lateralisation for the expression and perception of facial emotion in nonhuman primates. This author, indeed, confirmed that the patterns of hemispheric asymmetry that characterise the human brain are also evident in other primate species. These findings are not necessarily at odds with Gainotti's model, but are more inconsistent with Craig's (2005, 2009) hypothesis. Gainotti, in fact, suggested that the prevalence of the right hemisphere in emotional processing may be consonant with the more primitive nonverbal functioning of this hemisphere, but did not maintain that this prevalence was a consequence of the development of language within the left hemisphere. It is, indeed, equally possible that the right hemisphere's specialisation for emotion processing may have emerged early in primate evolution and that the development of language within the left side of the brain may be in part related to this right hemisphere specialisation for emotional functions. On the other hand, if the lateralisation of emotional functions was due (as Craig suggests), to an asymmetrical central representation of homeostatic activities, the same asymmetry of homeostatic afferent pathways subsuming these representations should also be observed in nonhuman primates showing a similar lateralisation of emotional functions. Craig (2005), however, acknowledges that this is not the case and this could raise some problems for his model.

References

Andersson S, Finset A. Heart rate and skin conductance reactivity to brief psychological stress in brain-injured patients. J Psychosom Res. 1998;44:645–56.

Barron SA, Rogovski Z, Hemli J. Autonomic consequences of cerebral hemisphere infarction. Stroke. 1994;25:113–6.

Bartolomeo P, Chokron S. Orienting of attention in left unilateral neglect. Neurosci Biobehav Rev. 2002;26:217–34.

Caltagirone C, Zoccolotti P, Originale G, Daniele A, Mammucari A. Autonomic reactivity and facial expression of emotions in brain-damaged patients. In: Gainotti G, Caltagirone C, editors. Emotions and the dual brain. Heidelberg: Springer; 1989. p. 204–21.

Cannon WB. The James-Lange theory of emotion: a critical examination and an alternative theory. Am J Psychol. 1927;39:106–24.

Colivicchi F, Bassi A, Santini M, Caltagirone C. Cardiac autonomic derangement and arrhythmias in right-sided stroke with insular involvement. Stroke. 2004;35:2094–8.

Craig AD. How do you feel? Interoception: the sense of the physiological condition of the body. Nat Rev Neurosci. 2002;3:655–66.

Craig AD. Forebrain emotional asymmetry: a neuroanatomical basis? Trends Cogn Sci. 2005;9:566–71.

Craig AD. How do you feel—now? The anterior insula and human awareness. Nat Rev Neurosci. 2009;10:59–70.

Craig AD. The sentient self. Brain Struct Funct. 2010;214:563–77.

Craig AD. Significance of the insula for the evolution of human awareness of feelings from the body. Ann N Y Acad Sci. 2011;1225:72–82.

Craig AD, Chen K, Bandy D, Reiman EM. Thermosensory activation of insular cortex. Nature. Neurosci. 2000;3:184–90.

Ekman P. Expression and the nature of emotion. In: Scherer K, Ekman P, editors. Approachs to emotion. Hillsdale, NJ: Erlbaum; 1984. p. 319–44.

Frijda NH. The emotions. Cambridge: Cambridge University Press; 1986.

Frijda NH. Emotions, cognitive structures and action tendency. Cogn Emot. 1987;1:115–43.

Gainotti G. Lateralization of brain mechanisms underlying automatic and controlled forms of spatial orienting of attention. Neurosci Biobehav Rev. 1996;20:617–22.

Gainotti G. Emotions, unconscious processes and the right hemisphere. Neuro-psychoanalysis. 2005;7:71–81.

Gainotti G. Face familiarity feelings, the right temporal lobe and the possibile underlying neural mechanisms. Brain Res Rev. 2007a;56:214–35.

Gainotti G. Different patterns of famous people recognition disorders in patients with right and left anterior temporal lesions: a systematic review. Neuropsychologia. 2007b;45:1591–607.

Gainotti G. What the study of voice recognition in normal subjects and brain-damaged patients tells us about models of familiar people recognition. Neuropsychologia. 2011;49:2273–82.

Gainotti G. The format of conceptual representations disrupted in semantic dementia: a position paper. Cortex. 2012;48:521–9.

Gainotti G. Is the difference between right and left ATLs due to the distinction between general and social cognition or between verbal and non-verbal representations? Neurosci Biobehav Rev. 2015;51:296–312.

Gainotti G. Emotions and the right hemisphere: can new data clarify old models? Neuroscientist. 2019a;25:258–70.

Gainotti G. The role of the right hemisphere in emotional and behavioral disorders of patients with frontotemporal lobar degeneration: an updated review. Front Aging Neurosci. 2019b; https://doi.org/10.3389/fnagi.2019.00055.

Guo CC, Sturm VE, Zhou J, Gennatas ED, Trujillo AJ, Hua AY, et al. Dominant hemisphere lateralization of cortical parasympathetic control as revealed by frontotemporal dementia. Proc Natl Acad Sci U S A. 2016;113:E2430–9.

Hanamori T, Kunitake T, Kato K, Kannan H. Responses of neurons in the insular cortex to gustatory, visceral, and nociceptive stimuli in rats. J Neurophysiol. 1998;79:2535–45.

Harmon-Jones E, Allen JJB. Anger and frontal brain activity: EEG asymmetry consistent with approach motivation despite negative affective valence. J Pers Soc Psychol. 1998;74:1310–6.

Harmon-Jones E, Sigelman J. State anger and prefrontal brain activity: evidence that insult-related relative left-prefrontal activation is associated with experienced anger and aggression. J Pers Soc Psychol. 2001;80:797–803.

Hewig J, Hagemann D, Seifert J, Naumann E, Bartussek D. On the selective relation of frontal cortical asymmetry and anger-out versus anger-control. J Pers Soc Psychol. 2004;87:926–39.

Làdavas E, Cimatti D, Del Pesce M, Tozzi G. Emotional evaluation with and without conscious stimulus identifications: evidence from a split-brain patient. Cogn Emot. 1993;7:95–114.

LeDoux J. The emotional brain. New York: Simon and Schuster; 1996.

Lindell AK. Continuities in emotion lateralization in human and non-human primates. Front Hum Neurosci. 2013;7:464. https://doi.org/10.3389/fnhum.2013.00464. eCollection 2013.

Morris JS, Ohman A, Dolan RJ, Rowland D, Young AW, Calder AJ, et al. Conscious and unconscious emotional learning in the human amygdala. Nature. 1998;393:467–70.

Morris JS, Ohman A, Dolan RJ. A subcortical pathway to the right amygdala mediating "unseen" fear. PNAS. 1999;96:1680–5.

Naver HK, Blomstrand C, Wallin G. Reduced heart rate variability after right-sided stroke. Stroke. 1996;27:247–51.

Oppenheimer SM, Gelb A, Girvin JP, Hachinski VC. Cardiovascular effects of human insular cortex stimulation. Neurology. 1992;42:1727–32.

Panksepp J. Affective consciousness in animals: perspectives on dimensional and primary process emotion approaches. Proc R Soc Lond B Biol Sci. 2010;277:2905–7.

Russell JA, Barrett LF. Core affect, prototypical emotional episodes, and other things called emotion: dissecting the elephant. J Pers Soc Psychol. 1999;76:805–19.

Shi CJ, Cassell MD. Cortical, thalamic, and amygdaloid connections of the anterior and posterior insular cortices. J Comp Neurol. 1998;399:440–68.

Snowden JS, Thompson JC, Neary D. Knowledge of famous faces and names in semantic dementia. Brain. 2004;127:860–72.

Spence S, Shapiro D, Zaidel E. The role of the right hemisphere in the physiological and cognitive components of emotional processing. Psychophysiology. 1996;33:112–22.

Sturm VE, Sible IJ, Datta S, Hua AY, Perry DC, Kramer JH, et al. Resting parasympathetic dysfunction predicts prosocial helping deficits in behavioral variant frontotemporal dementia. Cortex. 2018a;109:141–55.

Sturm VE, Brown JA, Hua AY, Lwi SJ, Zhou J, Kurth F, et al. Network architecture underlying basal autonomic outflow: evidence from Frontotemporal dementia. J Neurosci. 2018b;38:8943–55.

Tokgozoglu SL, Batur MK, Topçuoglu MA, Saribas O, Kes S, Oto A. Effects of stroke localization on cardiac autonomic balance and sudden death. Stroke. 1999;30:1307–11.

Toller G, Brown J, Sollberger M, Shdo SM, Bouvet L, Sukhanov P, et al. Individual differences in socioemotional sensitivity are an index of salience network function. Cortex. 2018;103:211–23.

Toller G, Yang WFZ, Brown JA, Ranasinghe KG, Shdo SM, Kramer JH, et al. Divergent patterns of loss of interpersonal warmth in frontotemporal dementia syndromes are predicted by altered intrinsic network connectivity. Neuroimage Clin. 2019;22:101729. https://doi.org/10.1016/j.nicl.2019.101729.

Vuilleumier P, Mohr C, Valenza N, Wetzel C, Landis T. Hyperfamiliarity for unknown faces after left lateral temporo-occipital venous infarction: a double dissociation with prosopagnosia. Brain. 2003;126:889–907.

Wallez C, Vauclair J. Human (Homo sapiens) and baboon (Papio papio) chimeric face processing: right-hemisphere involvement. J Comp Psychol. 2013;127:237–44.

Wittling W. Brain asymmetry in the control of autonomic-physiologic activity. In: Davidson RJ, Hugdahl K, editors. Brain asymmetry. Cambridge: MIT Press; 1995. p. 305–57.

Wittling W, Block A, Genzel S, Schweiger E. Hemisphere asymmetry in parasympathetic control of the heart. Neuropsychologia. 1998;36:461–8.

Woollams AM, Patterson K. Cognitive consequences of the left-right asymmetry of atrophy in semantic dementia. Cortex. 2018;107:64–77.

Yokoyama K, Jennings R, Ackles P, Hood BS, Boller F. Lack of heart rate changes during attention-demanding tasks after right hemisphere lesions. Neurology. 1987;37:624–30.

Zoccolotti P, Caltagirone C, Benedetti N, Gainotti G. Perturbation des réponses végétatives aux stimuli émotionnels au cours des lésions hémisphériques unilatérales. L'Encéphale. 1986;12:263–8.

Some Psychopathological Implications of Hemispheric Asymmetries for Emotions

7

Contents

We have seen in Sect. 4.6 of Chap. 4 that models which assume, respectively, a general dominance of the right hemisphere for emotional functions ('right hemisphere hypothesis') and a different hemispheric specialisation for negative and positive emotions ('valence hypothesis') allow us to make different predictions about the lateralisation of various aspects of psychopathology. On one hand, the 'valence hypothesis' might suggest that disruption of the left-sided structures subserving positive emotions could cause a major depression, whereas disorganisation of the right-sided structures subsuming negative emotions could elicit a manic syndrome. On the other hand, the 'right hemisphere hypothesis' might suggest that various aspects of psychopathology could be subtended by right hemisphere dysfunction, just as different kinds of abnormal emotional reactions are shown by right brain-damaged patients. In the same section of Chap. 4, we saw that predictions based on the 'valence hypothesis' could be supported by the observation that a major depression is usually found in stroke patients with left

© Springer Nature Switzerland AG 2020
G. Gainotti, *Emotions and the Right Side of the Brain*,
https://doi.org/10.1007/978-3-030-34090-2_7

frontal lesions whereas mania secondary to cerebrovascular lesions is mainly observed in patients with right brain lesions. We also saw, however, that these findings provoked strong debates and that different predictions based on the 'right hemisphere hypothesis' are also supported by a large amount of data. Since a detailed discussion of these complex and controversial issues could not be made in that section, but required a full chapter, we intend to develop this subject here by taking into account separately the problems: (a) of major post-stroke depression and of secondary mania and (b) of the large array of psychopathological conditions in which a prominent role of the right hemisphere has been reported or hypothesised.

7.1   The Hypothesis Which Assumes That Major Post-stroke Depression and Secondary Mania Are Caused Respectively by Left and Right Hemispheric Lesions

7.1.1   The Relationships Between Major Post-stroke Depression and Left Frontal Lesions

Post-stroke depression (PSD) is one of the most frequent neuropsychiatric consequences of stroke, which affects almost 30% of stroke survivors (Hackett et al. 2014) and can greatly influence the prognosis, not only 'quoad valetudinem' but also 'quoad vitam'. Moreover, PSD has a negative impact on functional outcome and rehabilitation results (Gainotti et al. 2001; Amaricai and Poenaru 2016). The results of studies on the frequency, determinants and consequences of PSD have been, however, very variable. Therefore, controversies about its incidence, its (biological or psychological) determinants, its consequences and its treatment have been frequent since the 1980s, when Robinson and coworkers (e.g. Robinson et al. 1983, 1984; Lipsey et al. 1986; Starkstein et al. 1988; Starkstein and Robinson 1989) published an important series of papers on affective disorders of stroke patients and proposed an influential model of post-stroke depression (PSD) that was consistent with the predictions of the 'valence hypothesis'. According to this model, it should be possible to distinguish, on the basis of the DSM-III diagnostic criteria (American Psychiatric Association 1980), two different forms of PSD: a major form, considered as an endogenous (psychotic) depression, and a minor form, considered as a neurotic or reactive depression. The major form should be provoked by left frontal lesions, which could disrupt lateralised monoaminergic pathways that connect the brainstem to the neocortex (Robinson et al. 1984); the minor form should be considered as a reactive form of depression, with no specific anatomo-clinical correlates. Even if the neurochemical mechanism proposed by Robinson and coworkers cannot be easily reconciled with the valence hypothesis the relationship between major PSD and left hemisphere damage was confirmed by other authors (e.g. Braun et al. 1999; Vataja et al. 2001) and their description of the symptomatological and neuroanatomical aspects of major PSD was very consistent with this model. Moreover, this consistency was further increased by the observation, made by the same authors (e.g. Robinson et al. 1988; Starkstein et al. 1990) that a right hemisphere lesion is often observed in the rare instances of secondary mania resulting from brain injury.

Robinson's model of PSD was very influential, but raised important methodological and anatomo-clinical objections. From the methodological point of view, it was argued that the criteria suggested by the DSM-III for making a diagnosis of major depression was not valid for patients with a symptomatic organic form of depression. In fact, according to the DSM-III a diagnosis of major depression can be made if, in addition to depressed mood, the patient presents at least five symptoms out of a series of eight; these include sleep disorders, loss of energy and lack of concentration, which, in a stroke patient, can be caused by a brain lesion per se and not by major depression. Gainotti et al. (1997a) checked the phenomenological equivalence between major post-stroke and endogenous depression (Lipsey et al. 1986) in a study in which they compared the main symptoms of patients classified as having major depression of vascular and endogenous origins on the basis of the DSM-III diagnostic criteria. This comparison was made with the post-stroke depression rating scale (PSDRS), which was specifically constructed to evaluate depression in stroke patients. The main features of the PSDRS are reported in Box 7.1:

Box 7.1 Main Features of the Post-stroke Depression Rating Scale
The PSDRS is a rating scale that is completed and scored by a professional (i.e. neurologist, psychologist or psychiatrist) examiner following a semi-structured patient interview.

The scale consists of 10 sections, each of which aims to evaluate a specific aspect of the emotional, affective and vegetative disorders of stroke patients. The 10 sections take into account the following different components of the anxious-depressive disorders of patients with PSD: (1) depressed mood, (2) feelings of guilt, (3) thoughts of death and/or suicide, (4) vegetative (sleep and appetite) disorders, (5) apathy and loss of interest, (6) anxiety (psychic and somatic anxiety, psychomotor agitation), (7) catastrophic reactions, (8) hyperemotionalism, (9) anhedonia (i.e. an inability to enjoy pleasant experiences) and (10) diurnal mood variations.

For each section (with the exception of the last one), scores range from 0 (corresponding to a normal state) to 5 (corresponding to severe disturbance).

In section 10 (diurnal mood variations), scores range between a negative pole (−2), corresponding to an 'unmotivated' prevalence of depression in the early morning and a positive pole (+2) corresponding to a 'motivated' prevalence of depression in situations that stress handicaps and disabilities. This section was introduced to evaluate whether the occurrence of depressive disorders in PSD patients is mainly due to biological, disease-related factors or to psychological and psychosocial factors. Furthermore, to better distinguish biological (unmotivated) from psychological (motivated) depressive disorders, in sections 1, 2 and 3 patients are requested to say whether their bad mood, guilt feelings and thoughts of death are related to aspects of their actual condition or are independent from them.

Therefore, the sum of the scores obtained in sections 1 to 9 can be considered as the 'global PSDRS score'; it allows evaluating the prevalence and severity of anxious-depressive disorders of stroke patients with a score ranging between 0 and 45 points. On the other hand, scores obtained by stroke patients in each section provide a symptomatological profile that can be matched to the profile shown by patients with major depressive disorder; this allows making inferences concerning the similarity or heterogeneity of the underlying mechanisms.

It was predicted that if the major form of PSD must be considered as a form of endogenous depression, as claimed by Robinson and coworkers, then the symptomatological profile obtained on the PSDRS by patients with major PSD should be very similar to that obtained by patients with (major) endogenous depression. In particular, both groups should obtain particularly high scores on the 'non-motivational' aspects of depression. On the contrary, the more motivated aspects should prevail in patients with minor PSD, considered by Robinson and coworkers as a form of neurotic/reactive depression. Neither of these predictions was confirmed by results of the study. In fact, (a) the symptomatological profiles of patients with major PSD were more similar to those of patients with minor PSD than to profiles of patients with endogenous depression; (b) the unmotivated aspects of depression (such as its prevalence early in the morning) prevailed in patients with endogenous depression, whereas the motivated aspects (such as its prevalence in situations that stressed the patient's disabilities) were in the foreground in patients with both major and minor PSD. From the anatomo-clinical point of view, most studies which checked the relationships between major PSD and left hemisphere (or left frontal) lesions (e.g. Agrell and Dehlin 1994; Andersen et al. 1995; Gainotti et al. 1997b, 1999; Singh et al. 1998; Carson et al. 2000; Bhogal et al. 2004; Hadidi et al. 2009; Hackett et al. 2014; Nickel and Thomalla 2017) failed to confirm the Robinson's model of PSD. On the contrary, a recent systematic review by Wei et al. (2015) not only failed to support the hypothesis of a link between the left hemisphere and increased risk of PSD, but also found a significant association between right hemisphere stroke and the incidence of depression in subacute post-stroke periods.

7.1.2   The Role of the Right Hemisphere in Secondary Mania After Brain Injury

Robinson et al. (1988) reviewed the literature concerning depression and mania caused by brain damage and underlined two main differences between these pathological conditions. The first concerned their incidence because, in contrast to the frequent occurrence of depression after ischemic brain lesions, mania had rarely been reported after brain damage. The second regarded the side of the lesion, because a review of anecdotal case reports by Cummings and Mendez (1984) suggested that the lesions of secondary manic patients are primarily in the right hemisphere.

In an effort to discover the factors that determine whether brain injury will lead to depression or mania, Robinson et al. (1988) compared a consecutive series of secondary mania patients with a similar series of patients who had major PSD; they confirmed that patients who develop secondary mania after brain injury have a greater frequency of injury to right hemisphere limbic areas, whereas patients with major PSD have injury primarily in the left frontal cortex and basal ganglia. The relations between secondary mania and right limbic lesions were confirmed by Starkstein et al. (1990) in another clinical neuroradiological and PET study of eight patients who developed a manic episode after brain injury.

The first weakening of the links between predictions based on the 'valence hypothesis' and results of anatomo-clinical investigations was, however, observed in another investigation in which Starkstein et al. (1991) compared patients who developed mania after a brain lesion with those who developed a bipolar affective disorder (i.e. mania and depression). In both patients' groups the lesions prevailed in the right hemisphere; however, they mainly concerned subcortical structures in bipolar patients and cortical structures (the right orbitofrontal and basotemporal cortices) in patients with unipolar mania. These results led Starkstein et al. (1991) to suggest that subcortical and cortical lesions of the right hemisphere might produce different neurochemical and/or remote metabolic brain changes that could underlie the production of either a bipolar disease or a unipolar mania. This interpretation, which assumes that a right brain lesion does not in itself cause mania, was subsequently developed by Starkstein and Robinson (1997), Mimura et al. (2005) and Braun et al. (2008). Starkstein and Robinson (1997) proposed that secondary mania due to orbitofrontal and basotemporal right hemisphere lesions might be primarily related to the release of left hemisphere influence, causing mental and behavioural disinhibition, rather than to a shift of mood toward a euphoric state. This hypothesis was confirmed by Mimura et al.'s (2005) description of a patient with recurrent depression who had undergone a cerebral infarction followed by post-stroke mania and in whom the comparison between pre- and post-stroke cerebral blood flow demonstrated a bilateral hypoperfusion in the frontal and the anterior cingulate cortices. The authors concluded that the patient's post-stroke mental state represented mania with disinhibition and anosognosia, rather than euphoria, and regarded it as part of a frontal lobe syndrome. Braun et al. (2008) also claimed that the manic symptoms resulting from right-sided lesions are due to a contralesional release phenomenon, i.e. activation of the left hemisphere after a right hemisphere lesion.

7.2   Psychopathological Conditions in Which a Prominent Role of the Right Hemisphere Has Been Hypothesised

In Sect. 4.6 of Chap. 4, we said that the 'right hemisphere hypothesis' might suggest that different kinds of psychopathology are subtended by right hemisphere dysfunction, just as different kinds of abnormal emotional reactions (observed in right brain-damaged patients) were considered by Gainotti (1972) as due to major involvement of the right hemisphere in emotional functions. We also said that the psychopathological conditions supporting these predictions concern, on one side, the lateral distribution of psychogenic pain and conversion reactions and, on the other side, the anatomical correlates of delusional reduplication and misidentification. Each of these psychopathological conditions will, therefore, be discussed separately in the next sections of this chapter.

7.2.1 Lateral Distribution of Psychogenic Pain and Conversion Reactions

Since the work of well-known French psychiatrists of the nineteenth century, such as Briquet (1859), it has been noted that in hysterical patients conversion reactions occur more frequently on the left than on the right side of the body. A similar asymmetry was also observed several years later by Halliday (1937) in patients with psychogenic pain. Drawing on the observation of Purves-Stewart (1924) that hysterical hyper-aesthesia is usually left-sided, but becomes right-sided in left-handers, Ferenczi (1926) interpreted this asymmetry as due to a lower level of activity on the left side of the body. Stern (1977) showed, however, that both right- and left-handed hysterical patients experience a higher proportion of motor and sensory symptoms on the left side of the body. This led Galin (1974), Galin et al. (1977) and Merskey and Watson (1979) to maintain that the left-sided preponderance of pain and conversion reactions is determined by the right hemisphere dominance for emotional experience. More recently, however, the left lateralisation of psychogenic pain and conversion reactions has become more controversial. Some authors (e.g. Lempert et al. 1990; Min and Lee 1997; Roelofs et al. 2000; Stone et al. 2002) have, in fact, claimed that there is insufficient support for lateralisation theories in conversion disorders, whereas other authors (e.g. Marshall et al. 1997; Devinsky et al. 2001; Perez et al. 2012) have tried to clarify the underlying pathophysiological mechanisms. Thus, Marshall et al. (1997) investigated the functional anatomy of a left-sided hysterical paralysis, studying changes in regional cerebral blood flow when the patient was asked to move the paralysed limbs. They showed that in this condition the patient failed to activate the right primary motor cortex, but significantly activated the ipsilateral orbitofrontal and anterior cingulate cortices. Marshall et al. (1997) concluded that these two areas could have inhibited consciously willed effects on the right primary motor cortex when the patient tried to move her left leg. On the other hand, Perez et al. (2012) discussed neural and clinical parallels between lesional unawareness disorders and unilateral motor and somatosensory conversion disorders, emphasising functional neuroimaging/disease correlates. They suggested that dysfunction of a functional-unawareness neurobiological framework, mediated by the right hemisphere anterior cingulate and posterior parietal cortices, might have a significant role in the neurobiology of the conversion disorder.

7.2.2   Neuroanatomical Correlates of Delusional Reduplication and Misidentification

Content specific delusions, involving the misidentification or the reduplication of persons, places or objects constitute a second area of psychopathology in which a critical role of the right hemisphere was clearly shown by the study of lesion laterality. In delusional misidentification syndromes (DMS) patients consistently misidentify persons, places or events (Feinberg and Roane 2005), whereas in reduplicative misidentification syndromes (RMS) they double the misidentified entity, expressing the subjective conviction that a place, person or event is duplicated. The most common form of misidentification is the Capgras syndrome (CS), described by Capgras and Reboul-Lachaux 1923), in which the patient maintains that a person well known to him (often a family member) has been replaced by an impostor or by a double and that there are two versions of this person. Another rare type of misidentification is the Fregoli syndrome (Courbon and Fail 1927), in which a person holds that different people are in fact a single person who changes appearance or is in disguise. On the other hand, patients with RMS believe that a place (or room or building) has been duplicated, existing in at least two locations simultaneously; this belief is termed reduplicative paramnesia (RP). According to Gainotti (1975), in this form of RMS the reduplicated version is usually closer to home than the hospital where the symptom is observed, due to the patient's need to deny his disease by replacing information pointing to illness with expressions suggesting normal health and efficiency. Several authors have shown that a high proportion of delusions involving misidentification and reduplication have a neurological basis and that lesions of the frontal lobes and the right hemisphere are critical for their development. Thus, Feinberg and Shapiro (1989) reviewed published cases of CS and RMS in whom brain dysfunction was implicated and found that in both conditions a lesion restricted to the right hemisphere was more common than a lesion restricted to the left hemisphere. Similar results were obtained by Murai et al. (1997), using a standard a questionnaire to elucidate the prevalence of RMS among patients with focal brain damage and by Forstl et al. (1991) who reviewed diverse groups of misidentification and reduplication syndromes and found right-sided abnormalities in 19 out of 20 patients with focal lesions on brain CT scans. More recently, similar investigations were conducted by Feinberg et al. (2005) in patients with delusional misidentification syndromes, by Politis and Loane (2012) in patients with reduplicative paramnesia and by Ardila (2018), reviewing psychiatric disorders associated with acquired brain pathology. Feinberg et al. (2005) reviewed a series of previously published patients who demonstrated persistent misidentifications or reduplications of either the Capgras or Fregoli type in the context of focal brain illness and identified 29 applicable reports of DMS. When cases with bilateral lesions were excluded, there were 14 (48.3%) cases of right hemisphere damage only and no cases of left hemisphere damage only. Politis and Loane (2012) made a systematic review of case reports, clinical studies and post-mortem studies that focussed on or referred to RP and found that right and bifrontal lesions were common factors in RP presentation and Ardila (2018) confirmed that somatoparaphrenia and delusional misidentification syndromes are usually associated with right hemisphere or bilateral lesions. Similar results were obtained by Gurin and Blum (2017) in a review of structural pathology in patients with delusional syndromes. The association between content-specific delusions and damage to the fronto-temporal limbic structures of the right hemisphere was also confirmed by Staff et al. (1999) using HMPAO–SPET to study the neural correlates of delusions in Alzheimer's disease patients. Several interpretations, based on specific functions of the right hemisphere, were advanced to account for the special role played by this side of the brain in content-specific delusions. Thus, Feinberg and Shapiro (1989) and Feinberg and Keenan (2005) drew on data showing that the right hemisphere has a key role in producing the experience of familiarity and suggested that delusional misidentifications might be caused by a defect of familiarity judgements. In fact, Ellis and Young (1990) suggested that patients with Capgras syndrome might have overt recognition of faces without the appropriate emotional reaction because they have a lesion in a 'dorsal route' that runs between the visual cortex and the limbic system through the inferior parietal lobule; and lesioning of the latter could allow for explicit recognition without the feeling of familiarity. On the other hand, Feinberg and Roane (2005) proposed that patients who have delusional misidentification and reduplication suffer from a disturbance of self and self-related functions; i.e. the right hemisphere might be dominant for the self, and right hemisphere damage might result in a disorder of ego boundaries and ego functions, which could explain why delusional misidentifications selectively concern aspects of the self or others of personal significance. All these interpretations suggest that the prevalent involvement of the right hemisphere in delusional misidentification syndromes might be caused by both the cognitive and the emotional characteristics of this hemisphere. The laterality of lesions producing content-specific delusions is, therefore, also consistent with predictions based on the 'right hemisphere hypothesis'. Also in line with the 'right hemisphere hypothesis'are investigations which have studied the neural correlates of psychotic disorders (delusions and schizophrenia-like psychosis) in stroke patients. Studying delusional ideations in patients with acute stroke, Kumral and Oztürk (2004) showed, in fact, that delusions are associated with right posterior temporoparietal lesions, whereas Devine et al. (2014) stressed the role of the right inferior frontal gyrus in the pathogenesis of post-stroke psychosis. Rather similar results were obtained investigating the anatomical locus of lesion in stroke patients with schizophrenia-like psychosis. Studying atypical (schizophreniform) psychosis in stroke patients, Rabins et al. (1991) showed, indeed, that preexisting subcortical atrophy and a right hemisphere lesion location are risk factors for developing schizophreniform disorder following stroke. These data were confirmed by Stangeland et al. (2018) in a systematic review of post-stroke psychosis defined by the presence of hallucinations or delusions, because these authors found that in these patients lesions typically involved right hemisphere structures.

Overall, the psychopathological data surveyed in this chapter show a clear relationship between right hemisphere lesions and some kinds of affective disturbances, such as mania and bipolar manic-depressive disorders (e.g. Starkstein et al. 1991). The same holds for content-specific delusions, such as Capgras syndrome (e.g. Feinberg et al. 2005) and reduplicative paramnesia (e.g. Politis and Loane 2012) or general psychotic disorders such as delusions and schizophrenia like psychosis observed in stroke patients (Stangeland et al. 2018). These data, therefore, suggest a relationship between the right hemisphere and various aspects of psychopathology and are more consistent with predictions based on the 'right hemisphere hypothesis' than with those based on the 'valence hypothesis'.

References

Agrell B, Dehlin O. Depression in stroke patients with left and right hemisphere lesions. A study in geriatric rehabilitation in-patients. Aging (Milano). 1994;6:49–56.

Amaricai E, Poenaru DV. The post-stroke depression and its impact onfunctioning in young and adult stroke patients of a rehabilitation unit. J Ment Health. 2016;25:137–41.

American Psychiatric Association. Diagnostic and statistical manual of mental disorders, 3rd ed (DSM-III). Washington, DC: American Psychiatric Press; 1980.

Andersen G, Vestergaard K, Ingemann-Nielsen M, Lauritzen L. Risk factors for post-stroke depression. Acta Psychiatr Scand. 1995;92:193–8.

Ardila A. Psychiatric disorders associated with acquired brain pathology. Appl Neuropsychol Adult. 2018;26:1–7. https://doi.org/10.1080/23279095.2018.1463224.

Bhogal SK, Teasell R, Foley N, Speechley M. Lesion location and poststroke depression: systematic review of the methodological limitations in the literature. Stroke. 2004;35:794–802.

Braun CMJ, Larocque C, Daigneault S, Montour-Proulx I. Mania, pseudomania, depression, and pseudodepression resulting from focal unilateral cortical lesions. Neuropsychiatry Neuropsychol Behav Neurol. 1999;12:35–51.

Braun CMJ, Daigneault R, Gaudelet S, Guimond A. Diagnostic and statistical manual of mental disorders, fourth edition symptoms of mania: which one(s) result(s) more often from right than left hemisphere lesions? Compr Psychiatry. 2008;49:441–59.

Briquet P. Traité clinique et thérapeutique de l'hystérie. Paris: Baillière; 1859.

Capgras J, Reboul-Lachaux J. L'illusion des "sosies"dans un delire systematize. Bull Soc Clin Med Ment. 1923;11:6–16.

Carson AJ, MacHale S, Allen K, Lawrie SM, Dennis M, House A, et al. Depression after stroke and lesion location: a systematic review. Lancet. 2000;356:122–6.

Courbon P, Fail G. Syndrome "d'illusion de Fregoli" et schizophrenie. Ann Med Psychol (Paris). 1927;85:289–90.

Cummings JL, Mendez MF. Secondary mania with focal cerebrovascular lesions. Am J Psychiatry. 1984;141:1084–7.

Devine MJ, Bentley P, Jones B, Hotton G, Greenwood RJ, Jenkins IH, et al. The role of the right inferior frontal gyrus in the pathogenesis of post-stroke psychosis. J Neurol. 2014;261:600–3.

Devinsky O, Mesad S, Alper K. Nondominant hemisphere lesions and conversion nonepileptic seizures. J Neuropsychiatry Clin Neurosci. 2001;13(3):367–73.

Ellis HD, Young AW. Accounting for delusional misidentification. Br J Psychiatry. 1990;57:239–48.

Feinberg TE, Keenan JP. Where in the brain is the self ? Consc Cognit. 2005;14:661–78.

Feinberg TE, Roane DM. Delusional Misidentification. Psychiatr Clin N Am. 2005;28:665–83.

Feinberg TE, Shapiro RM. Misidentification–reduplication and the right hemisphere. Cognitive and Behav Neurol. 1989;2:39–48.

Feinberg TE, DeLuca J, Giacino JT, Roane DM, Solms M. Right hemisphere pathology and the self: delusional misidentification and reduplication. In: Feinberg TE, Keenan JP, editors. The lost self: pathologies of the brain and identity. New York: Oxford; 2005. p. 100–30.

Ferenczi S. An attempted explanation of some hysterical stigmata. In: Ferenczi S, editor. Further contributions to the theory and technique of psychoanalysis. London: Hogarth Press; 1926. p. 110–7.

Forstl H, Almeida OP, Owen AM, Burns A, Howard R. Psychiatric, neurological and medical aspects of misidentification syndromes: a review of 260 cases. Psychol Med. 1991;21:905–10.

Gainotti G. Emotional behavior and hemispheric side of the lesion. Cortex. 1972;8:41–55.

Gainotti G. Confabulation of denial in senile dementia. An experimental study Psychiatr Clin (Basel). 1975;8:99–108.

Gainotti G, Azzoni A, Gasparini F, Marra C, Razzano C. Relation of lesion location to verbal and nonverbal mood measures in stroke patients. Stroke. 1997a;28:2145–9.

Gainotti G, Azzoni A, Razzano C, Lanzillotta M, Marra C, Gasparini F. The post-stroke depression rating scale: a test specifically devised to investigate affective disorders of stroke patients. J Clin Exp Neuropsychol. 1997b;19:340–56.

Gainotti G, Azzoni A, Marra C. Frequency, phenomenology and anatomical-clinical correlates of major post-stroke depression. Br J Psychiatry. 1999;175:163–7.

Gainotti G, Antonucci G, Marra C, Paolucci S. Relation between depression after stroke, antidepressant therapy, and functional recovery. J Neurol Neurosurg Psychiatry. 2001;71:258–61.

Galin D. Implications for psychiatry of left and right cerebral specialization. Arch Gen Psychiatry. 1974;31:572–83.

Galin D, Diamond R, Braff D. Lateralization of conversion symptoms: more frequent on the left. Am J Psychiatry. 1977;134:578–58.

Gurin L, Blum S. Delusions and the right hemisphere: a review of the case for the right hemisphere as a mediator of reality-based belief. J Neuropsychiatry Clin Neurosci. 2017;29:225–35.

Hackett ML, Köhler S, O'Brien JT, Mead GE. Neuropsychiatric outcomes of stroke. Lancet Neurol. 2014;13:525–34.

Hadidi N, Treat-Jacobson DJ, Lindquist R. Poststroke depression and functional outcome: a critical review of literature. Heart Lung. 2009;38:151–62.

Halliday JL. Psychological factors in rheumatism: a preliminary study. Br Med J. 1937;1:264–9.

Kumral E, Oztürk O. Delusional state following acute stroke. Neurology. 2004;62:110–3.

Lempert T, Dieterich M, Huppert D, Brandt T. Psychogenic disorders in neurology: frequency and clinical spectrum. Acta Neurol Scand. 1990;82:335–40.

Lipsey JR, Spencer WC, Rabins PV, Robinson RG. Phenomenological comparison of poststroke depression and functional depression. Am J Psychiatry. 1986;143:527–9.

Marshall JC, Halligan PW, Fink GR, Wade DT, Frackowiak RS. The functional anatomy of a hysterical paralysis. Cognition. 1997;64:B1–8.

Merskey H, Watson GD. The lateralisation of pain. Pain. 1979;7:271–80.

Mimura M, Nakagome K, Hirashima N, Ishiwata H, Kamijima K, Shinozuka A, et al. Left fronto-temporal hyperperfusion in a patient with post-stroke mania. Psychiatry Res. 2005;139:263–7.

Min SK, Lee BO. Laterality in somatization. Psychosom Med. 1997;59:236–40.

Murai T, Toichi M, Sengoku A, Miyoshi K, Morimune S. Reduplicative paramnesia in patients with focal brain damage. Neuropsychiatry Neuropsychol Behav Neurol. 1997;10:190–6.

Nickel A, Thomalla G. Post-stroke depression: impact of lesion location and methodological limitations—a topical review. Front Neurol. 2017;8. https://doi.org/10.3389/fneur.2017.00498.

Perez DL, Barsky AJ, Daffner K, Silbersweig DA. Motor and somatosensory conversion disorder: a functional unawareness syndrome? J Neuropsychiatry Clin Neurosci. 2012;24:141–51.

Politis M, Loane C. Reduplicative Paramnesia: a review. Psychopathology. 2012;45:337–43.

Purves-Stewart J. The diagnosis of neural diseases. London: Bulter & Tomner; 1924.

Rabins PV, Starkstein SE, Robinson RG. Risk factors for developing atypical (schizophreniform) psychosis following stroke. J Neuropsychiatry Clin Neurosci. 1991;3:6–9.

Robinson RG, Kubos KL, Starr LB, Rao K, Price TR. Mood changes in stroke patients: relationship to lesion location. Compr Psychiatry. 1983;24:555–66.

Robinson RG, Kubos KL, Starr LB, Rao K, Price TR. Mood disorders in stroke patients. Importance of location of lesion. Brain. 1984;107:81–93.

Robinson RG, Boston JD, Starkstein SE, Price TR. Comparison of mania and depression after brain injury: causal factors. Am J Psychiatry. 1988;145:172–8.

Roelofs K, Naring GW, Moene FC, Hoogduin CA. The question of symptom lateralization in conversion disorder. J Psychosom Res. 2000;49:21–5.

Singh A, Herrmann N, Black SE. The importance of lesion location in poststroke depression: a critical review. Can J Psychiatr. 1998;43:921–7.

Staff RT, Shanks MF, Macintosh L, Pestell SJ, Gemmell HG, Venneri A. Delusions in Alzheimer's disease: SPET evidence of right hemispheric dysfunction. Cortex. 1999;35:549–60.

Stangeland H, Orgeta V, Bell V. Poststroke psychosis: a systematic review. J Neurol Neurosurg Psychiatry. 2018;89:879–85.

Starkstein SE, Robinson RG. Affective disorders and cerebral vascular disease. Br J Psychiatry. 1989;154:170–82.

Starkstein SE, Robinson RG. Mechanism of disinhibition after brain lesions. J Nerv Ment Dis. 1997;185:108–14.

Starkstein SE, Robinson RG, Price TR. Comparison of patients with and without poststroke major depression matched for size and location of lesion. Arch Gen Psychiatry. 1988;45:247–52.

Starkstein SE, Mayberg HS, Berthier ML, Fedoroff P, Price TR, Dannals RF, et al. Mania after brain injury: neuroradiological and metabolic findings. Ann Neurol. 1990;27:652–9.

Starkstein SE, Fedoroff P, Berthier ML, Robinson RG. Manic-depressive and pure manic states after brain lesions. Biol Psychiatry. 1991;29:149–58.

Stern DB. Handedness and the lateral distribution of conversion reactions. J Nerv Ment Dis. 1977;164:122–8.

Stone J, Zeman A, Sharpe M. Functional weakness and sensory disturbance. J Neurol Neurosurg Psychiatry. 2002;73:241–5.

Vataja R, Pohjasvaara T, Leppävuori A, Mäntylä R, Aronen HJ, Salonen O, et al. Magnetic resonance imaging correlates of depression after ischemic stroke. Arch Gen Psychiatry. 2001;58:925–31.

Wei N, Yong W, Li X, Zhou Y, Deng M, Zhu H, et al. Post-stroke depression and lesion location: a systematic review. J Neurol. 2015;262:81–90.

Concluding Remarks: Special Relations Between Emotional System and Sympathetic Activities

Contents

The privileged relations between the emotional system and sympathetic activities have been repeatedly stressed in different sections of this volume. In fact, two important points support these special relations. The first is that, according to many authors (e.g. Ekman 1984; Frijda 1986; Oatley and Johnson-Laird 1987; LeDoux 1996; Panksepp 1998, 2010), the emotional system is basically an emergency system. The second is that the main function of a sympathetic activation is to allow the organism to respond quickly and strongly to emergency situations (Cannon 1927; Karplus and Kreidl 1909, 1927; Bard 1928; see Wittling 1995 for a short review). The special links between emotions and sympathetic activities are, therefore, self-evident. The close relationships between emotional system and sympathetically mediated emergency situations are also implicitly acknowledged by the 'negativity bias' (Carretie' et al. 2001; Rozin and Royzman 2001; Vaish et al. 2008) according to which organisms are more prepared to attend to and defend themselves from situations that signal danger/punishment than to attend to and approach situations that signal safety/reward.

All of this suggests that if the main components of the emotional system are mediated by right hemisphere structures, then the sympathetic section of the autonomic system, which is involved in emergency situations, should also be lateralised to this hemisphere. This suggestion is strongly supported by data obtained by Zoccolotti et al. (1986a, b), Yokoyama et al. (1987), Caltagirone et al. (1989), Wittling (1990, 1995); Andersson and Finset (1998) and Wittling et al. (1998), who studied the psychophysiological correlates of emotional activation in patients with unilateral brain lesions, and by Làdavas et al. (1993) and Spence et al. (1996), who studied physiological responses to the projection of emotional stimuli to the right

© Springer Nature Switzerland AG 2020
G. Gainotti, *Emotions and the Right Side of the Brain*,
https://doi.org/10.1007/978-3-030-34090-2_8

and left hemisphere of split-brain patients. This suggestion is also consistent with data obtained by Rosen et al. (1982), Zamrini et al. (1990), Oppenheimer et al. (1992), Sander and Klingelhofer (1995), Naver et al. (1996), Yoon et al. (1997), Tokgozoglu et al. (1999) and Oppenheimer (2006), who investigated hemispheric asymmetries for autonomic heart control (see Sect. 4.4 of this volume).

Less clear and more open to methodological objections are studies which have suggested an association between high parasympathetic activities and positive emotions. A general objection to this association could be that different kinds of positive emotions have been distinguished and that they could be characterised by qualitatively distinct profiles of autonomic activation. In fact, Shiota et al. (2011) and Duarte and Pinto-Gouveia (2017) noted that, since different positive emotions might have evolved for different purposes, different positive emotions might be differently associated with vegetative functioning Support for this suggestion can be found in Kreibig's (2010) review of experimental investigations that studied the relations between emotional stimuli and peripheral physiological responding in healthy individuals. This author found considerable autonomic response specificity when considering subtypes of distinct emotions. Even more consistent with Duarte and Pinto-Gouveia's (2017) statement were results obtained by Shiota et al. (2011), who assessed people's sympathetic and parasympathetic activations to five kinds of positive emotions (anticipatory enthusiasm, attachment love, nurturing love, amusement and awe). These authors showed that these different emotions are characterised by qualitatively distinct profiles of autonomic activation, thus providing evidence for the existence of autonomically distinct forms of positive emotions. These results clearly suggest caution in suggesting a crucial role of parasympathetic activities in positive emotions.

Apart from these methodological problems, it can be acknowledged that, since the parasympathetic section of the autonomic system is usually considered as a sort of control mechanism, aimed at counterbalancing the costs of the sympathetic response and restoring the organism's energetic resources, then there might also be some reason to expect weak left hemisphere lateralisation of parasympathetic activities. No real conflict could, therefore, exist between the two neurobiological models proposed by Gainotti (2005, 2007, 2012, 2019) and by Craig (2005, 2009, 2010, 2011), even though the former stresses the right hemisphere dominance for emotions and the latter the different hemispheric representation of sympathetic and parasympathetic homeostatic activities. In any case, we have seen in Sect. 4.4 of Chap. 4 and Sect. 6.3 of Chap. 6 that sympathetic activities are strongly lateralised to the right hemisphere, whereas parasympathetic activities are only weakly lateralised to the left hemisphere. We can, therefore, conclude that this pattern of lateralisation of sympathetic and parasympathetic activities is consistent with a model which assumes the existence of a special link between the right hemisphere, sympathetic activities and the most prototypical forms of basic emotions; by contrast, the relationships between the left hemisphere, parasympathetic activities and group-oriented positive forms of social emotions are less conclusive.

References

Andersson S, Finset A. Heart rate and skin conductance reactivity to brief psychological stress in brain-injured patients. J Psychosom Res. 1998;44:645–56.

Bard P. A diencephalic mechanism for the expression of rage with special reference to the sympathetic nervous system. Am J Physiology. 1928;84:490–515.

Caltagirone C, Zoccolotti P, Originale G, Daniele A, Mammucari A. Autonomic reactivity and facial expression of emotions in brain-damaged patients. In: Gainotti G, Caltagirone C, editors. Emotions and the dual brain. Heidelberg: Springer; 1989. p. 204–21.

Cannon WB. The James-Lange theory of emotion: a critical examination and an alternative theory. Am J Psychol. 1927;39:106–24.

Carretie' L, Mercado F, Tapia M, Hinojosa JA. Emotion, attention, and the 'negativity bias', studied through event-related potentials. Int J Psychophysiol. 2001;41:75–85.

Craig AD. Forebrain emotional asymmetry: a neuroanatomical basis? Trends Cogn Sci. 2005;9:566–71.

Craig AD. How do you feel—now? The anterior insula and human awareness. Nat Rev Neurosci. 2009;10:59–70.

Craig AD. The sentient self. Brain Struct Funct. 2010;214:563–77.

Craig AD. Significance of the insula for the evolution of human awareness of feelings from the body. Ann N Y Acad Sci. 2011;1225:72–82.

Duarte J, Pinto-Gouveia J. Positive affect and parasympathetic activity: evidence for a quadratic relationship between feeling safe and content and heart rate variability. Psychiatry Res. 2017;257:284–9.

Ekman P. Expression and the nature of emotion. In: Scherer K, Ekman P, editors. Approachs to emotion. Hillsdale, NJ: Erlbaum; 1984. p. 319–44.

Frijda NH. The emotions. Cambridge: Cambridge University Press; 1986.

Gainotti G. Emotions, unconscious processes and the right hemisphere. Neuro-psychoanalysis. 2005;7:71–81.

Gainotti G. Face familiarity feelings, the right temporal lobe and the possibile underlying neural mechanisms. Brain Res Rev. 2007;56:214–35.

Gainotti G. The format of conceptual representations disrupted in semantic dementia: a position paper. Cortex. 2012;48:521–9.

Gainotti G. Emotions and the right hemisphere: can new data clarify old models? Neuroscientist. 2019;25:258–70.

Karplus JP, Kreidl A. Gehirn und Sympathicus. I Zwischenhirnbasis und Hallsympathicus Pf Arch Gesam Physiol Men Thiere. 1909;129:138–44.

Karplus JP, Kreidl A. Gehirn und Sympathicus. VII Uber Beziehungen der Hypothalamuszentren zu Blutdruck und innerer Sekretion. Pf Arch Gesam Physiol Men Thiere. 1927;215:667–70.

Kreibig SD. Autonomic nervous system activity in emotion: a review. Biol Psychol. 2010;84:394–421.

Làdavas E, Cimatti D, Del Pesce M, Tozzi G. Emotional evaluation with and without conscious stimulus identifications: evidence from a split-brain patient. Cogn Emot. 1993;7:95–114.

LeDoux J. The emotional brain. New York: Simon and Schuster; 1996.

Naver HK, Blomstrand C, Wallin G. Reduced heart rate variability after right-sided stroke. Stroke. 1996;27:247–51.

Oatley K, Johnson-Laird P. Toward a cognitive theory of emotions. Cogn Emot. 1987;1:29–50.

Oppenheimer SM. Cerebrogenic cardiac arrhythmias: cortical lateralization and clinical significance. Clin Auton Res. 2006;16:6–11.

Oppenheimer SM, Gelb A, Girvin JP, Hachinski VC. Cardiovascular effects of human insular cortex stimulation. Neurology. 1992;42:1727–32.

Panksepp J. Affective neuroscience: the foundations of human and animal emotions. New York: Oxford University Press; 1998.

Panksepp J. Affective consciousness in animals: perspectives on dimensional and primary process emotion approaches. Proc R Soc Lond B Biol Sci. 2010;277:2905–7.

Rosen AD, Gur RC, Sussman N, Gur RE, Hurtig H. Hemispheric asymmetry in the control of heart rate. Abstr Social Neurosci. 1982;8:917.

Rozin P, Royzman EB. Negativity bias, negativity dominance, and contagion. Personal Soc Psychol Rev. 2001;5:296–320.

Sander D, Klingelhofer J. Changes of circadian blood pressure patterns and cardiovascular parameters indicate lateralization of sympathetic activation following hemispheric brain infarction. J Neurol. 1995;242:313–8.

Shiota MN, Neufeld SL, Yeung WH, Moser SE, Perea EF. Feeling good: autonomic nervous system responding in five positive emotions. Emotion. 2011;11:1368–78.

Spence S, Shapiro D, Zaidel E. The role of the right hemisphere in the physiological and cognitive components of emotional processing. Psychophysiology. 1996;33:112–22.

Tokgozoglu SL, Batur MK, Topçuoglu MA, Saribas O, Kes S, Oto A. Effects of stroke localization on cardiac autonomic balance and sudden death. Stroke. 1999;30:1307–11.

Vaish A, Grossmann T, Woodward A. Not all emotions are created equal: the negativity bias in socialemotional development. Psychol Bull. 2008;134:383–403.

Wittling W. Psychophysiological correlates of human brain asymmetry: blood pressure changes during lateralized presentation of an emotionally laden film. Neuropsychologia. 1990;28:457–70.

Wittling W. Brain asymmetry in the control of autonomic-physiologic activity. In: Asymmetry B, editor. Davidson RJ, Hugdahl K. Cambridge: MIT Press; 1995. p. 305–57.

Wittling W, Block A, Schweiger E, Genzel S. Hemisphere asymmetry in sympathetic control of the human myocardium. Brain Cogn. 1998;38:17–35.

Yokoyama K, Jennings R, Ackles P, Hood BS, Boller F. Lack of heart rate changes during attention-demanding tasks after right hemisphere lesions. Neurology. 1987;37:624–30.

Yoon BW, Morillo CA, Cechetto DF, Hachinski V. Cerebral hemispheric lateralization in cardial autonomic control. Arch Neurol. 1997;54:741–4.

Zamrini EY, Meador KJ, Loring DW, Nichols FT, Lee GP, Figueroa RE et al. Unilateral cerebral inactivation produces differential left/right heart rate responses. Neurology 1990;40:1408–11.

Zoccolotti P, Caltagirone C, Benedetti N, Gainotti G. Perturbation des réponses végétatives aux stimuli émotionnels au cours des lésions hémisphériques unilatérales. L'Encéphale. 1986a;12:263–8.

Zoccolotti P, Caltagirone C, Benedetti N, Gainotti G. Perturbation des réponses végétatives aux stimuli émotionnels au cours des lésions hémisphériques unilatérales. L'Encéphale. 1986b;12:263–8.